ASSESSMENT ITEM LISTING
FOR INSIDE THE
RESTLESS EARTH

HOLT, RINEHART AND WINSTON

A Harcourt Classroom Education Company

Austin · New York · Orlando · Atlanta · San Francisco · Boston · Dallas · Toronto · London

Copyright © by Holt, Rinehart and Winston

All rights reserved. No part of this publication may be reproduced or transmitted in any
form or by any means, electronic or mechanical, including photocopy, recording, or any
information storage and retrieval system, without permission in writing from the publisher.

Teachers using HOLT SCIENCE AND TECHNOLOGY may photocopy complete pages in
sufficient quantities for classroom use only and not for resale.

Art and Photo Credits

All work, unless otherwise noted, contributed by Holt, Rinehart and Winston.

Front cover: Frans Lanting/Minden Pictures; (owl on cover, title page) Kim Taylor/Bruce
Coleman, Inc.

Printed in the United States of America

ISBN 0-03-065514-5

1 2 3 4 5 6 082 05 04 03 02 01

CONTENTS

Copyright © by Holt, Rinehart and Winston. All rights reserved.

Copyright © by Holt, Rinehart and Winston. All rights reserved.

Introduction

The *Holt Science and Technology* Test Generator and *Assessment Item Listing*
The *Holt Science and Technology* Test Generator consists of a comprehensive bank of test items and the ExamView® Pro 3.0 software, which enables you to produce your own tests based on the items in the Test Generator and items you create yourself. Both Macintosh® and Windows® versions of the Test Generator are included on the *Holt Science and Technology* One-Stop Planner with Test Generator. Directions on pp. vi–vii of this book explain how to install the program on your computer. This *Assessment Item Listing* is a printout of all the test items in the *Holt Science and Technology* Test Generator.

ExamView Software
ExamView enables you to quickly create printed and on-line tests. You can enter your own questions in a variety of formats, including true/false, multiple choice, completion, problem, short answer, and essay. The program also allows you to customize the content and appearance of the tests you create.

Test Items
The *Holt Science and Technology* Test Generator contains a file of test items for each chapter of the textbook. Each item is correlated to the chapter objectives in the textbook and by difficulty level.

Item Codes
As you browse through this *Assessment Item Listing*, you will see that all test items of the same type appear under an identifying head. Each item is coded to assist you with item selection. Following is an explanation of the codes.

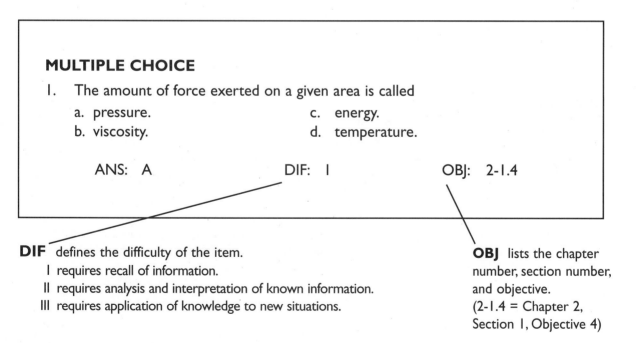

MULTIPLE CHOICE

1. The amount of force exerted on a given area is called
    a. pressure.
    b. viscosity.
    c. energy.
    d. temperature.

    ANS: A   DIF: I   OBJ: 2-1.4

DIF defines the difficulty of the item.
   I requires recall of information.
   II requires analysis and interpretation of known information.
   III requires application of knowledge to new situations.

OBJ lists the chapter number, section number, and objective.
(2-1.4 = Chapter 2, Section 1, Objective 4)

Copyright © by Holt, Rinehart and Winston. All rights reserved.

INSTALLATION AND STARTUP

The Test Generator is provided on the One-Stop Planner. The Test Generator includes ExamView and all of the questions for the corresponding textbook. ExamView includes three components: Test Builder, Question Bank Editor, and Test Player. The Test Builder includes options to create, edit, print, and save tests. The Question Bank Editor lets you create or edit question banks. The Test Player is a separate program that your students can use to take on-line* (computerized or LAN-based) tests. Please refer to the ExamView User's Guide on the One-Stop Planner for complete instructions.

Before you can use the Test Generator, you must install ExamView and the test banks on your hard drive. The system requirements, installation instructions, and startup procedures are provided below.

SYSTEM REQUIREMENTS

To use ExamView, your computer must meet or exceed the following hardware requirements:

Windows®
- Pentium processor
- Windows 95®, Windows 98®, Windows 2000® (or a more recent version)
- color monitor (VGA-compatible)
- CD-ROM and/or high-density floppy disk drive
- hard drive with at least 7 MB space available
- 8 MB available memory (16 MB memory recommended)
- an Internet connection (if you wish to access the Internet testing features)*

Macintosh®
- PowerPC processor, 100 MHz
- System 7.5 (or a more recent version)
- color monitor (VGA-compatible)
- CD-ROM and/or high-density floppy disk drive
- hard drive with at least 7 MB space available
- 8 MB available memory (16 MB memory recommended)
- an Internet connection with System 8.6 or a more recent version (if you wish to access the Internet testing features)*

* You can use the Test Player to host tests on your personal or school Web site or local area network (LAN) at no additional charge. The ExamView Web site's Internet test-hosting service must be purchased separately. Visit www.examview.com to learn more.

Copyright © by Holt, Rinehart and Winston. All rights reserved.

INSTALLATION

Instructions for installing ExamView from the CD-ROM:

Windows®
Step 1
Turn on your computer.
Step 2
Insert the One-Stop Planner into the CD-ROM drive.
Step 3
Click the Start button on the taskbar, and choose the Run option.
Step 4
In the Open box, type "d:\setup.exe" (substitute the letter for your drive if it is not d:) and click OK.
Step 5
Follow the prompts on the screen to complete the installation process.

Macintosh®
Step 1
Turn on your computer.
Step 2
Insert the One-Stop Planner into the CD-ROM drive. When the CD-ROM icon appears on the desktop, double-click the icon.
Step 3
Double-click the ExamView Pro Installer icon.
Step 4
Follow the prompts on the screen to complete the installation process.

Instructions for installing ExamView from the Main Menu of the One-Stop Planner (Macintosh® or Windows®):

Follow steps 1 and 2 from above.
Step 3
Double-click One-Stop.pdf. (If you do not have Adobe Acrobat® Reader installed on your computer, install it before proceeding by clicking Reader Installer.)
Step 4
To advance to the Main Menu, click anywhere on the title screen.
Step 5
Click the Test Generator button.
Step 6
Click the appropriate Install ExamView button.
Step 7
Follow the prompts on the screen to complete the installation process.

Copyright © by Holt, Rinehart and Winston. All rights reserved.

GETTING STARTED

After you complete the installation process, follow these instructions to start ExamView. See the ExamView User's Guide on the One-Stop Planner for further instructions on the program's options for creating a test and editing a question bank.

Startup Instructions

Step 1
Turn on the computer.

Step 2
Windows®: Click the Start button on the taskbar. Highlight the Programs menu, and locate the ExamView Pro Test Generator folder. Select the ExamView Pro option to start the software.
Macintosh®: Locate and open the ExamView Pro folder. Double-click the ExamView Pro icon.

Step 3
The first time you run the software, you will be prompted to enter your name, school/institution name, and city/state. You are now ready to begin using ExamView.

Step 4
Each time you start ExamView, the Startup menu appears. Choose one of the options shown.

Step 5
Use ExamView to create a test or edit questions in a question bank.

Technical Support

If you have any questions about the Test Generator, call the Holt, Rinehart and Winston technical support line at 1-800-323-9239, Monday through Friday, 7:00 A.M. to 6:00 P.M., Central Standard Time. You can contact the Technical Support Center on the Internet at http://www.hrwtechsupport.com or by e-mail at tsc@hrwtechsupport.com.

Copyright © by Holt, Rinehart and Winston. All rights reserved.

MULTIPLE CHOICE

1. On Mohs' hardness scale, which of the following minerals is harder than quartz?
   a. talc
   b. apatite
   c. gypsum
   d. topaz

   ANS: D   DIF: I   OBJ: 1-2.1

2. A mineral's streak
   a. is more reliable than color in identifying a mineral.
   b. reveals the mineral's specific gravity.
   c. is the same as a luster test.
   d. reveals the mineral's crystal structure.

   ANS: A   DIF: I   OBJ: 1-2.1

3. Which of the following terms is NOT used to describe a mineral's luster?
   a. pearly
   b. waxy
   c. dull
   d. hexagonal

   ANS: D   DIF: I   OBJ: 1-2.2

4. Which of the following is considered a special property that applies to only a few minerals?
   a. color
   b. luster
   c. streak
   d. magnetism

   ANS: D   DIF: I   OBJ: 1-2.2

5. Which of the following physical properties can be expressed in numbers?
   a. luster
   b. hardness
   c. color
   d. reaction to acid

   ANS: B   DIF: I   OBJ: 1-2.1

6. Which of the following minerals would scratch fluorite?
   a. talc
   b. quartz
   c. gypsum
   d. calcite

   ANS: B   DIF: I   OBJ: 1-2.1

7. If the atoms of a substance are arranged in a specific, repetitive pattern, the substance is
   a. a compound.
   b. colorful.
   c. crystalline.
   d. easily scratched.

   ANS: C   DIF: I   OBJ: 1-1.1

Copyright © by Holt, Rinehart and Winston. All rights reserved.

8. The mineral calcite, $CaCO_3$, is an example of a
   a. carbonate.
   b. silicate.
   c. sulfate.
   d. halide.

   ANS: A   DIF: I   OBJ: 1-1.2

9. The nonsilicate minerals
   a. do not contain oxygen.
   b. are all rare substances.
   c. all display good cleavage.
   d. include native elements.

   ANS: D   DIF: I   OBJ: 1-1.2

10. In addition to reclamation, a good way to reduce the environmental effects of mining is to
    a. recycle materials.
    b. dig deeper mines.
    c. use more metals.
    d. flush mines with water.

    ANS: A   DIF: I   OBJ: 1-3.2

11. The color of the powder that a mineral leaves on a piece of white, unglazed porcelain is called the mineral's
    a. hardness.
    b. luster.
    c. streak.
    d. scratch.

    ANS: C   DIF: I   OBJ: 1-2.1

12. Which statement best describes minerals?
    a. Rocks are made of minerals.
    b. Minerals are made of rocks.
    c. Minerals can be solid, liquid or gas.
    d. Crystals made by people are minerals.

    ANS: A   DIF: I   OBJ: 1-1.1

13. Halite is a(n) _____ of sodium and chlorine.
    a. atom
    b. element
    c. compound
    d. electron

    ANS: C   DIF: I   OBJ: 1-1.1

14. Which of the following describes gems?
    a. minerals that are valued for their beauty rather than for usefulness
    b. mineral crystals that are attractive and rare
    c. minerals that are hard enough to be cut and polished
    d. all of the above

    ANS: D   DIF: I   OBJ: 1-3.2

15. Which of the following minerals would most likely form in metamorphic rock?
    a. calcite and dolomite
    b. gypsum and halite
    c. garnet, graphite, and talc
    d. topaz and tourmaline

    ANS: C   DIF: I   OBJ: 1-3.1

16. Which of the following factors is NOT important in the formation of minerals?
    a. heat
    b. volcanic activity
    c. presence of ground water
    d. wind

    ANS: D   DIF: I   OBJ: 1-3.1

17. The repeating inner structure of a mineral is often reflected in the _____ of the crystal.
    a. size
    b. shape
    c. color
    d. Both (b) and (c)

    ANS: B   DIF: I   OBJ: 1-1.1

18. Gold is an example of a(n)
    a. oxide mineral.
    b. sulfide mineral.
    c. carbonate mineral.
    d. native element.

    ANS: D   DIF: I   OBJ: 1-1.2

19. The most common classification of minerals is based on
    a. color.
    b. chemical composition.
    c. size.
    d. shape.

    ANS: B   DIF: I   OBJ: 1-1.2

20. The two most common elements in the Earth's crust are
    a. nitrogen and oxygen.
    b. carbon and silicon.
    c. silicon and oxygen.
    d. oxygen and carbon.

    ANS: C   DIF: I   OBJ: 1-1.2

21. _____ minerals make up less than ten percent of the Earth's crust.
    a. Nonsilicate
    b. Silicate
    c. Carbonate
    d. Sulfate

    ANS: A   DIF: I   OBJ: 1-1.2

22. _____ minerals are the most common silicate mineral group in the Earth's crust.
    a. Carbonate
    b. Feldspar
    c. Mica
    d. Quartz

    ANS: B   DIF: I   OBJ: 1-1.2

23. _____ is the basic building block of many rocks.
    a. Corundum
    b. Fluorite
    c. Quartz
    d. Galena

    ANS: C   DIF: I   OBJ: 1-1.2

Copyright © by Holt, Rinehart and Winston. All rights reserved.

24. _____ minerals make up about half the Earth's crust.
    a. Feldspar
    b. Carbonate
    c. Quartz
    d. Mica

    ANS: A   DIF: I   OBJ: 1-1.2

25. _____ minerals are shiny and soft, and they separate easily into sheets when they break.
    a. Quartz
    b. Mica
    c. Feldspar
    d. Oxide

    ANS: B   DIF: I   OBJ: 1-1.2

26. _____ minerals contain the elements silicon and oxygen along with aluminum, potassium, sodium, and calcium.
    a. Feldspar
    b. Quartz
    c. Mica
    d. Halide

    ANS: A   DIF: II   OBJ: 1-1.2

27. SiO_2 is the chemical formula for which of the following minerals?
    a. calcite
    b. halite
    c. quartz
    d. fluorite

    ANS: C   DIF: II   OBJ: 1-1.2

28. Which of the following is a variety of mica mineral?
    a. calcite
    b. biotite
    c. dolomite
    d. fluorite

    ANS: B   DIF: II   OBJ: 1-1.2

29. Which of the following groups of minerals is a component of granite?
    a. feldspar minerals
    b. mica minerals
    c. quartz minerals
    d. all of the above

    ANS: D   DIF: I   OBJ: 1-1.2

30. _____ are the main component of most rocks on the Earth's surface.
    a. Carbonate minerals
    b. Feldspar minerals
    c. Mica minerals
    d. Quartz minerals

    ANS: B   DIF: I   OBJ: 1-1.2

31. NaCl is the chemical formula for which of the following minerals?
    a. halite
    b. calcite
    c. quartz
    d. galena

    ANS: A   DIF: II   OBJ: 1-1.2

32. You are out for a walk and find a piece of granite. The dark crystals scattered throughout the granite are most likely
   a. mica.   c. calcite.
   b. feldspar.   d. quartz.

   ANS: A   DIF: II   OBJ: 1-1.2

33. Gold, silver, and copper are all examples of
   a. oxides.   c. native elements.
   b. sulfides.   d. carbonates.

   ANS: C   DIF: I   OBJ: 1-1.2

34. Minerals that contain combinations of carbon and oxygen in their chemical structure are called
   a. native elements.   c. oxides.
   b. carbonates.   d. sulfates.

   ANS: B   DIF: II   OBJ: 1-1.2

35. Minerals that are composed of only one element are called
   a. sulfates.   c. oxides.
   b. carbonates.   d. native elements.

   ANS: D   DIF: I   OBJ: 1-1.2

36. We use _____ minerals in cement, building stones, and fireworks.
   a. oxide   c. sulfide
   b. carbonate   d. sulfate

   ANS: B   DIF: I   OBJ: 1-1.2

37. _____ are compounds that form when an element, such as aluminum or iron, combines chemically with oxygen.
   a. Sulfides   c. Oxides
   b. Carbonates   d. Sulfates

   ANS: C   DIF: II   OBJ: 1-1.2

38. _____ are minerals that contain both sulfur and oxygen (SO_4).
   a. Sulfides   c. Oxides
   b. Carbonates   d. Sulfates

   ANS: D   DIF: II   OBJ: 1-1.2

39. _____ are minerals that contain one or more elements that are combined with sulfur.
   a. Sulfides   c. Oxides
   b. Halides   d. Sulfates

   ANS: A   DIF: II   OBJ: 1-1.2

Copyright © by Holt, Rinehart and Winston. All rights reserved.

40. Scientists have shown that certain migratory fish can sense magnetic fields because they have magnetite in their brain. The mineral magnetite is a(n)
   a. sulfide.   c. oxide.
   b. carbonate.   d. sulfate.

   ANS: C   DIF: I   OBJ: 1-1.2

41. _____ are compounds that form when atoms of the elements fluorine (F), chlorine (Cl), iodine (I), or bromine (Br) combine with sodium (Na), potassium (K), or calcium (Ca).
   a. Sulfides   c. Oxides
   b. Halides   d. Sulfates

   ANS: B   DIF: II   OBJ: 1-1.2

42. The mineral galena (PbS) is an example of a(n)
   a. sulfide.   c. carbonate.
   b. oxide.   d. sulfate.

   ANS: A   DIF: II   OBJ: 1-1.2

43. The mineral gypsum ($CaSO_4 \cdot 2H_2O$) is an example of a
   a. carbonate.   c. native element.
   b. sulfide.   d. sulfate.

   ANS: D   DIF: II   OBJ: 1-1.2

44. Minerals are classified based on chemical composition into which two groups?
   a. silicates and nonsilicates   c. organic and inorganic
   b. carbons and noncarbons   d. crystalline and noncrystalline

   ANS: A   DIF: I   OBJ: 1-1.2

45. Which of the following minerals is a silicate mineral?
   a. calcite   c. halite
   b. fluorite   d. quartz

   ANS: D   DIF: I   OBJ: 1-1.2

46. Which of the following is NOT a property that is commonly used to identify minerals?
   a. color   c. density
   b. hardness   d. texture

   ANS: D   DIF: I   OBJ: 1-2.1

47. Color is sometimes useful in distinguishing quartz from amethyst. In its purest state, quartz is
   a. clear.   c. gold.
   b. rose.   d. purple.

   ANS: A   DIF: I   OBJ: 1-2.2

48. Because of factors such as weathering and impurities, ____ usually is NOT a reliable indicator of a mineral's identity.
    a. luster   c. color
    b. streak   d. hardness

    ANS: C   DIF: I   OBJ: 1-2.1

49. Which of the following terms is used to describe the nonmetallic luster of a mineral?
    a. resinous   c. vitreous
    b. waxy   d. all of the above

    ANS: D   DIF: I   OBJ: 1-2.1

50. Metallic, submetallic, and nonmetallic are descriptions of a mineral's
    a. streak.   c. color.
    b. luster.   d. cleavage.

    ANS: B   DIF: I   OBJ: 1-2.2

51. Different types of minerals break in different ways. Diamonds break
    a. in four different directions.   c. into distinct sheets.
    b. at 90° angles in three directions. d. None of the above

    ANS: A   DIF: I   OBJ: 1-2.3

52. When quartz breaks it forms a curved pattern called conchoidal
    a. luster.   c. fracture.
    b. cleavage.   d. hardness.

    ANS: C   DIF: I   OBJ: 1-2.2

53. Gem cutters take advantage of natural ____ to remove a flaws from diamonds and rubies.
    a. luster   c. fracture
    b. cleavage   d. hardness

    ANS: B   DIF: I   OBJ: 1-2.3

54. On Mohs' hardness scale, which of the following minerals is the softest?
    a. talc   c. corundum
    b. diamond   d. orthoclase

    ANS: A   DIF: I   OBJ: 1-2.1

55. Because ____ has a density of 1 g/cm^3, it is used as a reference point for other substances.
    a. oxygen   c. water
    b. gold   d. quartz

    ANS: C   DIF: I   OBJ: 1-2.1

Copyright © by Holt, Rinehart and Winston. All rights reserved.

56. You have a mineral that you are fairly certain is fluorite. Which test should help you determine if it really is fluorite?
    a. It should glow under ultraviolet light.
    b. It should attract iron.
    c. It should be detected by a Geiger counter.
    d. It should have a salty taste.

    ANS: A   DIF: I   OBJ: 1-2.2

57. The special property used to identify halite is
    a. radioactivity.   c. fluorescence.
    b. magnetism.   d. taste.

    ANS: D   DIF: I   OBJ: 1-2.2

58. Minerals that contain radium
    a. cause a double image.   c. can be detected by a Geiger counter.
    b. attract iron.   d. effervesce.

    ANS: C   DIF: I   OBJ: 1-2.2

59. You have a mineral that you are fairly certain is magnetite. Which test should show that it really is magnetite?
    a. It should glow under ultraviolet light.
    b. It should attract iron.
    c. It should be detected by a Geiger counter.
    d. It should have a salty taste.

    ANS: B   DIF: I   OBJ: 1-2.2

60. Which of the following tests could help you determine whether a mineral is calcite?
    a. It should effervesce when a drop of weak acid is placed on it.
    b. It should glow under ultraviolet light.
    c. It should cause a double image when a thin, clear piece is placed over an image.
    d. all of the above

    ANS: D   DIF: II   OBJ: 1-2.2

61. Gold and copper generally form
    a. on the bottom of lakes and seas.   c. in pegmatites.
    b. in hot-water solutions.   d. All of the above

    ANS: B   DIF: I   OBJ: 1-3.1

62. Many gems, such as topaz and tourmaline, form in
    a. pegmatites.   c. plutons.
    b. limestones.   d. metamorphic rocks.

    ANS: A   DIF: I   OBJ: 1-3.1

63. A body of magma that has cooled and solidified before reaching the surface is called a(n)
   a. pegmatite.
   b. evaporite.
   c. pluton.
   d. None of the above

   ANS: C   DIF: I   OBJ: 1-3.1

64. Which of the following is a type of surface mine?
   a. an open pit
   b. a strip mine
   c. a quarry
   d. all of the above

   ANS: D   DIF: I   OBJ: 1-3.2

65. Which of the following is a type of deep mine?
   a. a shaft
   b. a quarry
   c. an open pit
   d. a strip mine

   ANS: A   DIF: I   OBJ: 1-3.2

66. Mineral ores are
   a. renewable resources.
   b. nonrenewable resources.
   c. replaceable resources.
   d. available in infinite amounts.

   ANS: B   DIF: I   OBJ: 1-3.2

67. Gold classified as 18-karat is 18 parts gold and 6 parts another, similar metal. What is the percentage of pure gold in 18-karat gold?
   a. 18 percent
   b. 25 percent
   c. 33 percent
   d. 75 percent

   ANS: D   DIF: II   OBJ: 1-3.2

68. Copper is extracted from which of the following mineral ores?
   a. beryl
   b. chalcopyrite
   c. galena
   d. chromite

   ANS: B   DIF: I   OBJ: 1-3.2

69. What method is used to mine diamonds?
   a. deep mining
   b. surface mining
   c. underwater mining
   d. plateau mining

   ANS: A   DIF: I   OBJ: 1-3.2

70. Which of the following minerals is a natural magnet that attracts iron?
   a. galena
   b. pyrrhotite
   c. magnetite
   d. both (b) and (c)

   ANS: D   DIF: II   OBJ: 1-2.1

Copyright © by Holt, Rinehart and Winston. All rights reserved.

71. In which environment are small crystals formed due to slow cooling of hot magma beneath Earth's crust?
    a. limestones
    b. hot-water solutions
    c. plutons
    d. pegmatites

    ANS: C   DIF: I   OBJ: 1-3.1

72. Which of the following minerals generally forms from evaporating saltwater?
    a. garnet
    b. gypsum
    c. galena
    d. gold

    ANS: B   DIF: II   OBJ: 1-3.1

73. When ____ evaporates, minerals such as gypsum and halite are left behind.
    a. salt water
    b. fresh water
    c. a hot-water solution
    d. groundwater

    ANS: A   DIF: I   OBJ: 1-3.1

74. Limestone minerals such as calcite and dolomite that have been dissolved in and carried by surface water and groundwater crystallize
    a. in caverns.
    b. in the bottom of lakes and seas.
    c. underground.
    d. along shorelines.

    ANS: B   DIF: I   OBJ: 1-3.1

COMPLETION

1. A naturally formed, inorganic solid with a definite internal geometric structure is called a _____. (compound or mineral)

    ANS: mineral   DIF: I   OBJ: 1-1.1

2. _____ is the tendency of minerals to break along flat surfaces. (Cleavage or Fracture)

    ANS: Cleavage   DIF: I   OBJ: 1-2.1

3. Gold and silver are _____ because they are each composed of a single type of atom. (elements or compounds)

    ANS: elements   DIF: I   OBJ: 1-1.1

4. Aluminum comes from a mineral _____ called bauxite. (element or ore)

    ANS: ore   DIF: I   OBJ: 1-3.2

5. A material's _____ is defined as its mass divided by its volume. (fracture or density)

   ANS: density   DIF: I   OBJ: 1-2.1

6. The smallest part of an element that has all the properties of that element is called a(n) _____.

   ANS: atom   DIF: I   OBJ: 1-1.1

7. A substance made of two or more elements that have been chemically joined, or bonded together, is called a _____.

   ANS: compound   DIF: I   OBJ: 1-1.1

8. Minerals that contain a combination of silicon and oxygen are called _____ minerals.

   ANS: silicate   DIF: I   OBJ: 1-1.2

9. Minerals that do NOT contain a combination of silicon and oxygen are called _____ minerals.

   ANS: nonsilicate   DIF: I   OBJ: 1-1.2

10. _____ is a rock composed of various minerals, including feldspar, mica, and quartz.

    ANS: Granite   DIF: I   OBJ: 1-1.2

11. The way a surface reflects light is called _____.

    ANS: luster   DIF: I   OBJ: 1-2.1

12. The piece of unglazed porcelain that you rub a mineral against to help identify that mineral is called a(n) _____.

    ANS: streak plate   DIF: I   OBJ: 1-2.1

13. The tendency of some minerals to break unevenly along curved or irregular surfaces is called _____.

    ANS: fracture   DIF: I   OBJ: 1-2.1

14. _____ refers to a mineral's resistance to being scratched.

    ANS: Hardness   DIF: I   OBJ: 1-2.1

15. The ratio of an object's density to the density of water is called the object's
_____.

ANS: specific gravity   DIF: I   OBJ: 1-2.1

16. Geologists use the term _____ to describe a mineral deposit large enough and pure enough to be mined for a profit.

ANS: ore   DIF: I   OBJ: 1-3.2

17. Returning land to its original state after mining is completed is a process called
_____.

ANS: reclamation   DIF: I   OBJ: 1-3.2

18. When changes in pressure, temperature, or chemical makeup alter a rock,
_____ takes place.

ANS: metamorphism   DIF: I   OBJ: 1-3.1

SHORT ANSWER

For each pair of terms, explain the difference in their meaning.

1. mineral/atom

ANS:
A mineral is made up of a particular arrangement of different kinds of atoms.

DIF: I   OBJ: 1-1.1

2. silicate mineral/nonsilicate mineral

ANS:
Silicate minerals are made of silicon and oxygen compounds, while nonsilicate minerals are made of other compounds.

DIF: I   OBJ: 1-1.2

3. density/hardness

ANS:
The hardness of a mineral is its resistance to being scratched, while the density of a mineral is a measure of the amount of matter in a given space.

DIF: I   OBJ: 1-2.1

4. color/streak

ANS:
Streak is the color of a mineral in powdered form. The color of a mineral may change, but the mineral's streak is always the same.

DIF: I   OBJ: 1-2.1

5. element/compound

ANS:
Elements are made of only one kind of atom, while compounds are made of two or more elements that are chemically bonded.

DIF: I   OBJ: 1-1.1

6. fracture/cleavage

ANS:
If a mineral breaks along a curved or irregular surface, it has fracture. If a mineral breaks along flat surfaces, it has cleavage.

DIF: I   OBJ: 1-2.1

7. What are the differences between atoms, compounds, and minerals?

ANS:
Compounds are composed of two or more atoms of different elements that are chemically bonded. Minerals consist of atoms or compounds arranged in a crystalline structure.

DIF: I   OBJ: 1-1.1

8. Which two elements are most common in minerals?

ANS:
Oxygen and silicon are the most common elements in minerals.

DIF: I   OBJ: 1-1.2

9. How are silicate minerals different from nonsilicate minerals?

ANS:
Silicate minerals are made of combinations of the elements silicon and oxygen, and nonsilicate minerals are not.

DIF: I   OBJ: 1-1.2

10. Explain why each of the following is not considered a mineral: a cupcake, water, teeth, oxygen.

ANS:
A cupcake is not a mineral because it does not have a crystalline structure and it does not form in nature. Liquid water is not a mineral because it does not have a crystalline structure and it is a liquid, not a solid. (Water ice can be considered a mineral.) Teeth are not minerals because they are a living part of your body. Oxygen is not a mineral because oxygen atoms by themselves do not have a crystalline structure.

DIF: II   OBJ: 1-1.1

11. How do you determine a mineral's streak?

ANS:
Scrape the mineral across a ceramic streak plate. The color of the material that rubs off the mineral sample is the mineral's streak.

DIF: I   OBJ: 1-2.1

12. What is the difference between cleavage and fracture?

ANS:
If a mineral has cleavage, it breaks along flat surfaces. Fracture is the way a mineral breaks along curved or irregular surfaces.

DIF: I   OBJ: 1-2.2

13. How would you determine the hardness of an unidentified mineral sample?

ANS:
To determine the hardness of an unknown mineral sample, take a material of known hardness and try to scratch the unknown mineral with it. If the unknown mineral is scratched, try to scratch it with a material that has a lower hardness. Continue with this process until you know which materials are harder and which are softer than the unknown mineral sample. The hardness of the unknown mineral is between these two.

DIF: I   OBJ: 1-2.1

14. Suppose you have two minerals that have the same hardness. Which other mineral properties would you use to determine whether the samples are the same mineral?

ANS:
Answers will vary but should not include color.

DIF: II   OBJ: 1-2.1

Holt Science and Technology
Copyright © by Holt, Rinehart and Winston. All rights reserved.

15. Describe how minerals form underground.

ANS:
Answers will vary. Minerals form when magma cools and solidifies. Changing temperature and pressure conditions inside the Earth can cause the formation of new minerals from pre-existing rock. Minerals can also form underground when dissolved solids crystallize out of heated ground water.

DIF: I   OBJ: 1-3.1

16. What are the two main types of mining?

ANS:
The two main types of mining are surface mining and deep mining.

DIF: I   OBJ: 1-3.2

17. How does reclamation protect the environment around a mine?

ANS:
Reclamation reduces the harmful effects of mining by returning the land to its original state.

DIF: II   OBJ: 1-3.2

18. What is a mineral?

ANS:
A mineral is a naturally formed, inorganic solid with a crystalline structure.

DIF: I   OBJ: 1-1.1

19. What does a crystal's shape depend on?

ANS:
A crystal's shape depends on the arrangement of the atoms within the crystal.

DIF: I   OBJ: 1-1.1

20. Why is color not always a reliable way of identifying a mineral?

ANS:
Factors such as weathering and the inclusion of impurities can affect the mineral's color.

DIF: I   OBJ: 1-2.1

Copyright © by Holt, Rinehart and Winston. All rights reserved.

21. What property do minerals that glow under ultraviolet light display?

ANS:
They display the property of fluorescence.

DIF: I   OBJ: 1-2.2

22. Name three minerals that form in metamorphic rock.

ANS:
Possible answers: calcite, muscovite, chlorite, garnet, graphite, hematite, magnetite, and talc.

DIF: I   OBJ: 1-3.1

23. What is ore?

ANS:
Ore is the term geologists use to describe mineral deposits that are large enough and pure enough to be mined for profit.

DIF: I   OBJ: 1-3.2

24. How can mining cause water pollution?

ANS:
The waste products from a mine can introduce toxic concentrations of elements in rivers, lakes, and ground water.

DIF: I   OBJ: 1-3.2

25. Using no more than 25 words, define the term mineral.

ANS:
A mineral is a naturally occurring inorganic solid with a crystalline structure.

DIF: I   OBJ: 1-1.1

26. In one sentence, describe how density is used to identify a mineral.

ANS:
Each mineral has its own unique density.

DIF: I   OBJ: 1-2.1

Copyright © by Holt, Rinehart and Winston. All rights reserved.

27. What methods of mineral identification are the most reliable? Explain.

ANS:
Answers will vary. Cleavage, hardness, and density are very reliable because they can be measured and do not change. Color and fracture are less reliable.

DIF: I   OBJ: 1-2.1

28. Use the following terms to create a concept map: *minerals, oxides, nonsilicates, carbonates, silicates, hematite, calcite, quartz.*

ANS:

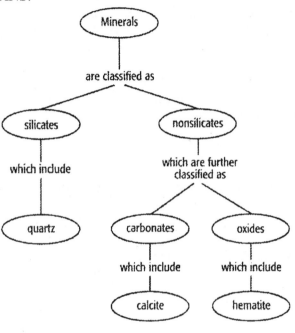

DIF: II   OBJ: 1-1.2

29. Suppose you have three rings, each with a different gem. One has a diamond, one has an amethyst (purple quartz), and one has a topaz. You mail the rings in a small box to your friend who lives five states away. When the box arrives at its destination, two of the gems are damaged. One gem, however, is damaged much worse than the other. What scientific reason can you give for the difference in damage?

ANS:
Each mineral has a different hardness. The hardest mineral was damaged the least. (The diamond will not be damaged, the topaz will be slightly damaged, and the amethyst will sustain the most damage.)

DIF: II   OBJ: 1-1.2

Copyright © by Holt, Rinehart and Winston. All rights reserved.

30. While trying to determine the identity of a mineral, you decide to do a streak test. You rub the mineral across the plate, but it does not leave a streak. Does this mean your test failed? Explain your answer.

ANS:
No; the test was actually successful. You learned that the unknown mineral has no streak and that it is harder than the streak plate. This clue will help you classify the mineral.

DIF: II   OBJ: 1-2.1

31. Imagine that you work at a jeweler's shop and someone brings in some "gold nuggets" that they want to sell. The person claims that an old prospector found the gold nuggets during the California gold rush. You are not sure if the nuggets are real gold. How would you decide whether to buy the nuggets? Which identification tests would help you decide the nuggets' identity?

ANS:
Students should suggest performing several tests to see whether the mineral is gold or not. Gold is very dense and very soft, so one would start with density and hardness tests.

DIF: II   OBJ: 1-2.1

32. Suppose that you find a mineral crystal that is as tall as you are. What kinds of environmental factors would cause such a crystal to form?

ANS:
Answers will vary. When magma contains a lot of hot fluids and cools slowly, very large crystals can grow.

DIF: II   OBJ: 1-3.1

33. Gold has a specific gravity of 19. Pyrite's specific gravity is 5. How much denser is gold than pyrite?

ANS:
Gold is $19 \div 5 = 3.8$ (three and four-fifths) times as dense as pyrite.

DIF: II   OBJ: 1-2.1

34. In a quartz crystal there is one silicon atom for every two oxygen atoms. That means that the ratio of silicon atoms to oxygen atoms is 1:2. If there were 8 million oxygen atoms in a sample of quartz, how many silicon atoms would there be?

ANS:
8 million $\div 2 = 4$ million silicon atoms

DIF: II   OBJ: 1-1.2

Copyright © by Holt, Rinehart and Winston. All rights reserved.

Imagine that you have a sample of feldspar and analyze it to find out what it is made of. The results of your analysis are shown below.

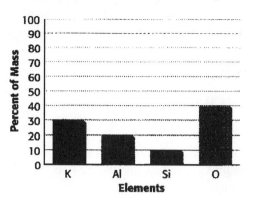

Composition of Orthoclase (Pink Feldspar)

35. Your sample consists of four elements. What percentage of each one is your sample made of?

ANS:
K: 30%
Al: 20%
Si: 10%
O: 40%

DIF: II   OBJ: 1-1.2

36. If your mineral sample has a mass of 10 g, how many grams of oxygen does it contain?

ANS:
4 g

DIF: II   OBJ: 1-1.2

37. Make a circle graph showing how much of each of the four elements the feldspar contains.

ANS:
Answers will vary.

DIF: II   OBJ: 1-1.2

38. Why is color not very useful for classifying minerals?

ANS:
The same mineral can exist in a wide variety of colors. This difference in color is caused by impurities in the mineral. Weathering can also change the mineral's color.

DIF: I   OBJ: 1-2.1

Copyright © by Holt, Rinehart and Winston. All rights reserved.

39. Describe two methods used to mine minerals.

ANS:
Surface mining is used to remove minerals at or near the Earth's surface. In deep mining, minerals are removed from deep within the Earth by the use of underground passageways.

DIF: I   OBJ: 1-3.2

40. How can mining harm the environment?

ANS:
Sample answer: Waste products can pollute both surface water and ground water. With surface mining, large areas of natural vegetation may be cleared, and that can disturb or destroy the habitats of plants and animals.

DIF: I   OBJ: 1-3.2

41. You are given two mineral samples. Sample A has a mass of 50 g and displaces 25 mL of water. Sample B has a mass of 114 g and displaces 54 mL of water. Which sample has the greater density? Show your work.

ANS:
density = mass ÷ volume
sample A: $D = 50$ g/25 cm^3 = 2 g/cm^3
sample B: $D = 114$ g/54 cm^3 = 2.1 g/cm^3
Sample B has the greater density.

DIF: III   OBJ: 1-2.1

42. Rebecca found a glimmering crystal that looked like a diamond, but Judy said it was only quartz. How could Rebecca use a crystal that she already knows is quartz to find out if her crystal is diamond or quartz?

ANS:
Sample answers: She could try to scratch a piece of quartz with her crystal. If her crystal scratches the quartz, it might be diamond. Rebecca could measure the mass and volume of her sample, then compare her crystal's specific gravity with that of quartz. She could also compare the lusters or crystal structures of the two crystals.

DIF: II   OBJ: 1-2.1

Holt Science and Technology
Copyright © by Holt, Rinehart and Winston. All rights reserved.

43. Use the table to answer the item that follows.

Comparative Hardness Scale

Hardness	Common material
2.5	fingernail
3	copper penny
5	steel knife blade

Bobby found a rock that he could scratch with his knife. The rock scratched a penny. Estimate the hardness of the rock.

ANS:
The rock has a hardness greater than three but less than five.

DIF: II   OBJ: 1-2.1

Copyright © by Holt, Rinehart and Winston. All rights reserved.

44. Use the following terms to complete the concept map below: *lakes and seas, cooling, chemical composition, evaporates, hot water solutions.*

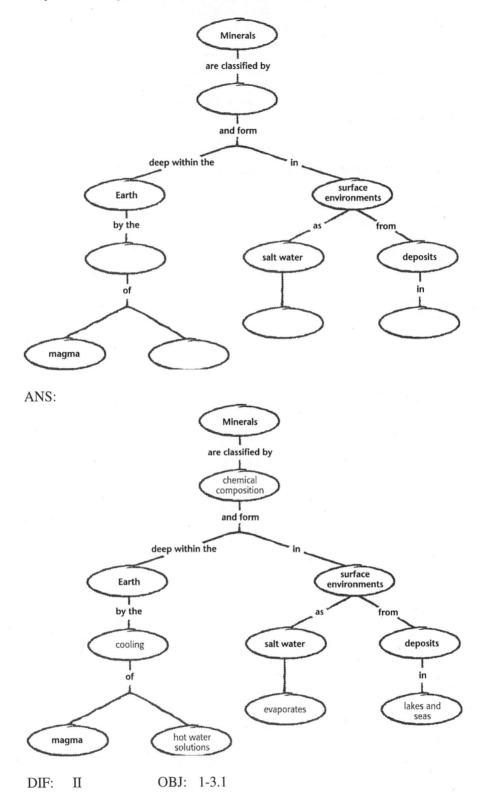

ANS:

DIF: II   OBJ: 1-3.1

MULTIPLE CHOICE

1. A type of rock that forms deep within the Earth when magma solidifies is called
   a. sedimentary.
   b. metamorphic.
   c. organic.
   d. igneous.

   ANS: D   DIF: I   OBJ: 2-1.2

2. A type of rock that forms under high temperature and pressure but is not exposed to enough heat to melt the rock is
   a. sedimentary.
   b. metamorphic.
   c. organic.
   d. igneous.

   ANS: B   DIF: I   OBJ: 2-1.2

3. After they are deposited, sediments, such as sand, are turned into sedimentary rock when they are compacted and
   a. cemented.
   b. metamorphosed.
   c. melted.
   d. weathered.

   ANS: A   DIF: I   OBJ: 2-3.1

4. An igneous rock with a coarse-grained texture forms when
   a. magma cools very slowly.
   b. magma cools very quickly.
   c. magma cools quickly, then slowly.
   d. magma cools slowly, then quickly.

   ANS: A   DIF: I   OBJ: 2-2.2

5. The layering that occurs in sedimentary rock is called
   a. foliation.
   b. ripple marks.
   c. stratification.
   d. compaction.

   ANS: C   DIF: I   OBJ: 2-3.1

6. An example of clastic sedimentary rock is
   a. obsidian.
   b. sandstone.
   c. gneiss.
   d. marble.

   ANS: B   DIF: I   OBJ: 2-3.1

7. A common sedimentary rock structure is
   a. a sill.
   b. a pluton.
   c. cross-bedding.
   d. a lava flow.

   ANS: C   DIF: I   OBJ: 2-3.1

Copyright © by Holt, Rinehart and Winston. All rights reserved.

8. An example of mafic igneous rock is
   a. granite.
   b. basalt.
   c. quartzite.
   d. pumice.

   ANS: B   DIF: I   OBJ: 2-2.2

9. Chemical sedimentary rock forms when
   a. magma cools and solidifies.
   b. minerals are twisted into a new arrangement.
   c. minerals crystallize from a solution.
   d. sand grains are cemented together.

   ANS: C   DIF: I   OBJ: 2-3.1

10. Which of the following is a foliated metamorphic rock?
    a. sandstone
    b. gneiss
    c. shale
    d. basalt

    ANS: B   DIF: I   OBJ: 2-4.3

11. Which of these causes the breakdown of rock?
    a. weathering
    b. cementing
    c. stratification
    d. extrusion

    ANS: A   DIF: I   OBJ: 2-1.2

12. Metamorphic rock is most likely to form when
    a. a volcano erupts explosively.
    b. limestone comes in contact with an igneous intrusion.
    c. erosion of weathered granite occurs very rapidly.
    d. silt particles are compacted and cemented.

    ANS: B   DIF: I   OBJ: 2-4.1

13. Phyllite is a rock formed from shale. What type of rock is phyllite?
    a. an igneous intrusive rock
    b. an igneous extrusive rock
    c. sedimentary rock
    d. metamorphic rock

    ANS: D   DIF: I   OBJ: 2-4.3

14. Which type of rock cooled very slowly?
    a. hard fossiliferous limestone
    b. coarse-grained granite
    c. loose sandstone
    d. fine-grained basalt

    ANS: B   DIF: I   OBJ: 2-2.2

15. The individual mineral grains in sandstone hold together because they have been
    a. cemented.
    b. melted.
    c. chemically changed.
    d. metamorphosed.

    ANS: A   DIF: I   OBJ: 2-3.1

Holt Science and Technology
Copyright © by Holt, Rinehart and Winston. All rights reserved.
24

16. Some scalpels used today are formed using
    a. flint.
    b. chert.
    c. obsidian.
    d. coal.

    ANS: C   DIF: I   OBJ: 2-1.1

17. Sedimentary rock is formed when grains of sediment undergo
    a. compaction and cementation.
    b. metamorphism.
    c. melting.
    d. solidification.

    ANS: A   DIF: I   OBJ: 2-1.2

18. Which of the following processes is NOT part of the rock cycle?
    a. metamorphism
    b. solidification
    c. erosion
    d. chiseling

    ANS: D   DIF: I   OBJ: 2-1.2

19. Which type of rock can weather and wear away to form sediment?
    a. igneous rock
    b. sedimentary rock
    c. metamorphic rock
    d. all of the above

    ANS: D   DIF: I   OBJ: 2-1.2

20. The weathering and wearing away of rock into sediment is called
    a. compaction.
    b. metamorphism.
    c. melting.
    d. erosion.

    ANS: D   DIF: I   OBJ: 2-1.2

21. Magma forms in Earth's
    a. lower crust.
    b. outer core.
    c. upper mantle.
    d. both (a) and (c)

    ANS: D   DIF: I   OBJ: 2-1.2

22. Earth scientists classify the three main types of rock into smaller groups according to
    a. composition and texture.
    b. density and hardness.
    c. color and luster.
    d. streak and cleavage.

    ANS: A   DIF: I   OBJ: 2-1.3

23. A piece of granite rock that you are studying is made of 30 percent quartz, 55 percent feldspar, and the rest biotite mica. What percentage of the rock is biotite mica?
    a. 5 percent
    b. 15 percent
    c. 25 percent
    d. 85 percent

    ANS: B   DIF: II   OBJ: 2-1.3

Copyright © by Holt, Rinehart and Winston. All rights reserved.

24. The sizes, shapes, and positions of the grains of a rock determine the rock's
    a. texture.   c. cleavage.
    b. luster.   d. streak.

    ANS: A   DIF: I   OBJ: 2-1.3

25. Fine-grained, medium-grained, and coarse-grained are descriptions of a rock's
    a. composition.   c. luster.
    b. texture.   d. streak.

    ANS: B   DIF: I   OBJ: 2-1.3

26. When magma or lava cools down enough, it solidifies to form
    a. sedimentary rock.   c. igneous rock.
    b. metamorphic rock.   d. all of the above

    ANS: C   DIF: I   OBJ: 2-2.1

27. At what temperatures do lava and magma solidify, or "freeze"?
    a. at 0°C   c. between 700°C and 1,250°C
    b. between 1,300°C and 2,000°C   d. none of the above

    ANS: C   DIF: I   OBJ: 2-2.1

28. Magma can form when
    a. rock is heated.   c. rock changes composition.
    b. pressure is released.   d. all of the above

    ANS: D   DIF: I   OBJ: 2-2.1

29. What can happen when a fluid, such as water or carbon dioxide, enters a rock that is close to its melting point?
    a. The fluid can combine with the rock and dissolve it.
    b. The fluid can lower the melting point of the rock enough to melt it.
    c. The fluid can raise the melting point of the rock enough to melt it.
    d. The fluid does not affect the melting point of the rock.

    ANS: B   DIF: I   OBJ: 2-2.1

30. Light-colored igneous rock generally has a _____ composition.
    a. felsic   c. nonfoliated
    b. foliated   d. mafic

    ANS: A   DIF: I   OBJ: 2-2.2

31. Dark-colored igneous rock generally has a _____ composition.
    a. felsic   c. nonfoliated
    b. foliated   d. mafic

    ANS: D   DIF: I   OBJ: 2-2.2

Copyright © by Holt, Rinehart and Winston. All rights reserved.

32. Granite is coarse-grained and generally contains light-colored minerals. It is an example of
    a. felsic igneous rock.
    b. mafic igneous rock.
    c. metamorphic rock.
    d. sedimentary rock.

    ANS: A   DIF: I   OBJ: 2-2.2

33. Fast-cooling lava will most likely form ____ igneous rock.
    a. fine-grained
    b. foliated
    c. nonfoliated
    d. coarse-grained

    ANS: A   DIF: I   OBJ: 2-2.2

34. Sills and laccoliths are ____ rock formations.
    a. foliated metamorphic
    b. nonfoliated metamorphic
    c. intrusive igneous
    d. extrusive igneous

    ANS: C   DIF: I   OBJ: 2-2.3

35. Which of the following is NOT an example of an intrusive igneous rock formation?
    a. a dike
    b. a lava plateau
    c. a pluton
    d. a sill

    ANS: B   DIF: I   OBJ: 2-2.3

36. Most volcanic rock is
    a. nonfoliated.
    b. intrusive.
    c. extrusive.
    d. foliated.

    ANS: C   DIF: I   OBJ: 2-2.3

37. Which of the following is NOT an example of an extrusive igneous rock?
    a. granite
    b. pumice
    c. obsidian
    d. basalt

    ANS: A   DIF: II   OBJ: 2-2.3

38. Which type of rock forms near the Earth's surface without the influence of intense heat and pressure?
    a. igneous rock
    b. sedimentary rock
    c. metamorphic rock
    d. all of the above

    ANS: B   DIF: I   OBJ: 2-3.1

39. Millions of years of erosion by the Colorado River have revealed the sedimentary rock ____ in the walls of the Grand Canyon.
    a. dikes
    b. sills
    c. strata
    d. batholiths

    ANS: C   DIF: I   OBJ: 2-3.1

Copyright © by Holt, Rinehart and Winston. All rights reserved.

40. Which of the following is the correct order of steps in a sedimentary rock cycle?
    a. weathering, erosion, deposition, compaction and cementation
    b. deposition, erosion, weathering, compaction and cementation
    c. compaction and cementation, deposition, erosion
    d. none of the above

    ANS: A   DIF: I   OBJ: 2-3.1

41. Which of the following is NOT an example of a clastic sedimentary rock?
    a. shale   c. sandstone
    b. slate   d. breccia

    ANS: B   DIF: I   OBJ: 2-3.1

42. Which of the following clastic sedimentary rocks is the most fine-grained?
    a. breccia   c. siltstone
    b. sandstone   d. shale

    ANS: D   DIF: I   OBJ: 2-3.1

43. Which of the following clastic sedimentary rocks is the most coarse-grained?
    a. conglomerate   c. siltstone
    b. sandstone   d. shale

    ANS: A   DIF: I   OBJ: 2-3.1

44. Rock salt (NaCl) is an example of a(n) _____ rock.
    a. intrusive igneous   c. chemical sedimentary
    b. clastic sedimentary   d. foliated metamorphic

    ANS: C   DIF: I   OBJ: 2-3.1

45. _____ is formed when calcium and carbonate become so concentrated in sea water that calcite crystallizes out.
    a. Marble   c. Fossiliferous limestone
    b. Chemical limestone   d. Shale

    ANS: B   DIF: I   OBJ: 2-3.1

46. Chemical limestone forms
    a. on the ocean floor.   c. on lava plateaus.
    b. in a magma chamber.   d. in fissures.

    ANS: A   DIF: I   OBJ: 2-3.1

47. The shells or skeletons of organisms that form fossiliferous limestone are composed of
    a. spongin.   c. cartilage.
    b. silicate.   d. calcium carbonate.

    ANS: D   DIF: I   OBJ: 2-3.1

Holt Science and Technology
Copyright © by Holt, Rinehart and Winston. All rights reserved.
28

48. As coral die, their skeletons accumulate on the ocean floor and are compacted and cemented into
    a. chemical limestone.
    c. fossiliferous limestone.
    b. shale.
    d. marble.

    ANS: C   DIF: I   OBJ: 2-3.1

49. Fossils are the remains or traces of plants and animals that have been preserved in ____ rock.
    a. granite
    c. sedimentary
    b. igneous
    d. metamorphic

    ANS: C   DIF: I   OBJ: 2-3.2

50. Most fossils come from animals that lived
    a. in the oceans.
    c. in tropical rain forests.
    b. on mountain tops.
    d. in deserts.

    ANS: A   DIF: I   OBJ: 2-3.2

51. Strata differ from one another depending on the
    a. kind of sediment they are composed of.
    b. color of their sediment.
    c. size of their sediment.
    d. All of the above

    ANS: D   DIF: I   OBJ: 2-3.2

52. When flowing water transports and deposits sediments, the sediments dry in
    a. strata.
    c. sills.
    b. caverns.
    d. plutons.

    ANS: A   DIF: I   OBJ: 2-3.1

53. When strata are slanted, they are called
    a. sills.
    c. cross-beds.
    b. batholiths.
    d. plutons.

    ANS: C   DIF: II   OBJ: 2-3.2

54. The word *metamorphic* comes from *meta-* and *morphos-* meaning
    a. "cooked rock."
    c. "cooled magma."
    b. "changed shape."
    d. "pressurized sediment."

    ANS: B   DIF: I   OBJ: 2-4.1

55. When rock undergoes metamorphism, the ____ of the rock changes.
    a. structure
    c. composition
    b. texture
    d. All of the above

    ANS: D   DIF: I   OBJ: 2-4.1

Copyright © by Holt, Rinehart and Winston. All rights reserved.

56. Which type of rock can change into metamorphic rock?
   a. igneous rock
   b. sedimentary rock
   c. metamorphic rock
   d. all of the above

   ANS: D   DIF: I   OBJ: 2-4.1

57. Most metamorphic change is caused by
   a. extreme heat.
   b. extreme cold.
   c. increased pressure.
   d. decreased pressure.

   ANS: C   DIF: I   OBJ: 2-4.1

58. At depths greater than 16 km below the Earth's surface, the pressure can be more than 4,000 times the pressure of the atmosphere. This can cause
   a. metamorphism.
   b. erosion.
   c. weathering.
   d. deposition.

   ANS: A   DIF: I   OBJ: 2-4.1

59. An example of a metamorphic rock is
   a. granite.
   b. garnet schist.
   c. fossiliferous limestone.
   d. basalt.

   ANS: B   DIF: I   OBJ: 2-4.1

60. The temperature at which metamorphism occurs ranges from
   a. 0°C to 30°C.
   b. 30°C to 50°C.
   c. 50°C to 1,000°C.
   d. 1,000°C and greater.

   ANS: C   DIF: I   OBJ: 2-4.1

61. At temperatures higher than 1,000°C, most rocks will
   a. solidify.
   b. melt.
   c. crystallize.
   d. bend.

   ANS: B   DIF: I   OBJ: 2-4.1

62. Which of the following statements does NOT describe metamorphism?
   a. Metamorphism melts rock.
   b. Metamorphism is caused by heat and/or pressure.
   c. Metamorphism generally occurs at depths greater than 2 km.
   d. Metamorphism causes rocks to appear cooked or squeezed.

   ANS: A   DIF: I   OBJ: 2-4.1

63. Contact metamorphism occurs
   a. near flowing water.
   b. underneath continental rock formations.
   c. next to magma.
   d. at the Earth's crust.

   ANS: C   DIF: I   OBJ: 2-4.1

Copyright © by Holt, Rinehart and Winston. All rights reserved.

64. Regional metamorphism usually occurs
    a. over small areas.
    b. underneath continental rock formations.
    c. over large areas.
    d. Both (b) and (c)

    ANS: D   DIF: I   OBJ: 2-4.1

65. Under increased pressure and heat, the minerals calcite, quartz, and hematite combine and
    recrystallize to form the mineral garnet. This is an example of
    a. metamorphism.   c. weathering.
    b. erosion.   d. deposition.

    ANS: A   DIF: I   OBJ: 2-4.2

66. Minerals that form when rock is metamorphosed
    a. may be minerals that occur only in metamorphic rock.
    b. can be used to estimate the temperature and pressure at which the rock metamorphosed.
    c. are more stable in their new environment.
    d. All of the above

    ANS: D   DIF: II   OBJ: 2-4.2

67. The ____ of metamorphic rock can be foliated or nonfoliated.
    a. cleavage   c. texture
    b. streak   d. luster

    ANS: C   DIF: I   OBJ: 2-4.3

68. Which of the following types of foliated metamorphic rock requires the least amount of heat and
    pressure to form?
    a. slate   c. schist
    b. phyllite   d. gneiss

    ANS: A   DIF: II   OBJ: 2-4.3

69. Which of the following types of foliated metamorphic rock requires the greatest amount of heat
    and pressure to form?
    a. slate   c. schist
    b. phyllite   d. gneiss

    ANS: D   DIF: II   OBJ: 2-4.3

70. What type of rock is shale?
    a. fine-grained, mafic igneous   c. clastic sedimentary
    b. fine-grained, foliated metamorphic   d. nonfoliated metamorphic

    ANS: C   DIF: II   OBJ: 2-3.1

Copyright © by Holt, Rinehart and Winston. All rights reserved.

71. What type of rock is slate?
   a. intrusive igneous
   b. fine-grained, foliated metamorphic
   c. clastic sedimentary
   d. nonfoliated metamorphic

   ANS: B   DIF: II   OBJ: 2-4.3

72. What type of rock is marble?
   a. intrusive igneous
   b. fine-grained, foliated metamorphic
   c. chemical sedimentary
   d. nonfoliated metamorphic

   ANS: D   DIF: II   OBJ: 2-4.3

73. Quartzite forms when _____ is metamorphosed.
   a. sandstone
   b. calcite
   c. hermatite
   d. granite

   ANS: A   DIF: II   OBJ: 2-4.3

74. Marble forms when _____ is metamorphosed.
   a. sandstone
   b. limestone
   c. phyllite
   d. granite

   ANS: B   DIF: II   OBJ: 2-4.3

75. Nonfoliated metamorphic rock will usually form _____ mineral from which it formed.
   a. smaller crystals of the same
   b. smaller crystals of a completely different
   c. larger crystals of the same
   d. larger crystals of a completely different

   ANS: C   DIF: II   OBJ: 2-4.3

76. Which of the following is used to classify metamorphic rock?
   a. clastic or chemical
   b. foliated or nonfoliated
   c. intrusive or extrusive
   d. felsic or mafic

   ANS: B   DIF: I   OBJ: 2-4.3

COMPLETION

1. _____ igneous rock is more likely to have coarse-grained texture than _____ igneous rock. (Extrusive/intrusive or Intrusive/extrusive)

   ANS: Intrusive/extrusive   DIF: I   OBJ: 2-2.3

2. _____ metamorphic rock texture consists of parallel alignment of mineral grains. (Foliated or Nonfoliated)

   ANS: Foliated   DIF: I   OBJ: 2-4.3

3. _____ sedimentary rock forms when grains of sand become cemented together. (Clastic or Chemical)

    ANS: Clastic   DIF: I   OBJ: 2-3.1

4. _____ cools quickly on the Earth's surface. (Lava or Magma)

    ANS: Lava   DIF: I   OBJ: 2-2.2

5. Strata are found in _____ rock. (igneous or sedimentary)

    ANS: sedimentary   DIF: I   OBJ: 2-3.2

6. Concrete is composed of rock particles bound together with cement. Thus, concrete is most like a _____ rock. (sedimentary or metamorphic)

    ANS: sedimentary   DIF: I   OBJ: 2-3.1

7. The _____ of a rock depends on the size of its crystals. (texture or composition)

    ANS: texture   DIF: I   OBJ: 2-1.3

8. Obsidian is an _____ igneous rock, which cools very rapidly on the Earth's surface. (intrusive or extrusive)

    ANS: extrusive   DIF: I   OBJ: 2-1.3

9. The cooling rate of magma affects the properties of _____ rock. (igneous or sedimentary)

    ANS: igneous   DIF: I   OBJ: 2-2.2

10. Slate is one of the _____ metamorphic rocks, which contain flat minerals. (foliated or nonfoliated)

    ANS: foliated   DIF: I   OBJ: 2-4.3

11. _____ is a solid mixture of crystals of one or more minerals.

    ANS: Rock   DIF: I   OBJ: 2-1.1

12. The _____ is the process by which one rock type changes into another.

    ANS: rock cycle   DIF: I   OBJ: 2-1.2

13. _____ is the hot liquid that forms when rock partially or completely melts.

    ANS: Magma   DIF: I   OBJ: 2-1.2

Copyright © by Holt, Rinehart and Winston. All rights reserved.

14. The minerals a rock is made of determine the rock's _____.

     ANS: composition   DIF: I   OBJ: 2-1.3

15. When magma cools beneath the Earth's surface, the resulting rock is called
     _____ igneous rock.

     ANS: intrusive   DIF: I   OBJ: 2-2.2

16. _____ are large, balloon-shaped intrusive formations that result when magma
     cools at great depths.

     ANS: Plutons   DIF: I   OBJ: 2-2.3

17. When lava erupts from a volcano, a formation called a _____ is made.

     ANS: lava flow   DIF: I   OBJ: 2-2.3

18. Sometimes lava erupts from long cracks in the Earth's surface called _____.

     ANS: fissures   DIF: I   OBJ: 2-2.3

19. When a large amount of lava flows out of a fissure, it can cover a vast area, forming a plain
     called a _____.

     ANS: lava plateau   DIF: I   OBJ: 2-2.3

20. _____ sedimentary rock forms when minerals crystallize out of a solution,
     such as sea water, to become rock.

     ANS: Chemical   DIF: I   OBJ: 2-3.1

21. If the temperature or pressure of a new environment is different from the one in which a rock is
     formed, the rock will undergo _____.

     ANS: metamorphism   DIF: I   OBJ: 2-4.1

22. _____ metamorphism only happens next to igneous intrusions.

     ANS: Contact   DIF: I   OBJ: 2-4.1

23. _____ metamorphism occurs when enormous pressure builds up in rock that is
     deeply buried under other rock formations, or when large pieces of the Earth's crust collide with
     each other.

     ANS: Regional   DIF: I   OBJ: 2-4.1

24. _____ metamorphic rock does not appear to have any regular pattern.

ANS: Nonfoliated   DIF: I   OBJ: 2-4.3

SHORT ANSWER

1. List two ways rock is important to humans today.

   ANS:
   Answers will vary.

   DIF: I   OBJ: 2-1.1

2. What are the three major rock types, and how can they change from one type to another?

   ANS:
   The three major rock types are igneous, sedimentary, and metamorphic. Igneous rock forms when magma cools and solidifies. Sedimentary rock forms when sediments are cemented and compacted together or when minerals crystallize out of sea water. Metamorphic rock forms when the texture or mineral composition of a preexisting rock is changed by heat or pressure.

   DIF: I   OBJ: 2-1.2

3. How is lava different from magma?

   ANS:
   Magma is a hot liquid that exists underground. Lava is magma that erupts and flows onto the Earth's surface.

   DIF: I   OBJ: 2-1.2

4. Explain the difference between texture and composition.

   ANS:
   The texture of a rock is determined by the sizes, shapes, and positions of the grains that make it up. The composition of a rock is determined by the kinds of minerals the rock is made of.

   DIF: II   OBJ: 2-1.3

5. What two properties are used to classify igneous rock?

   ANS:
   The two properties used to classify igneous rock are texture and color (mineral composition).

   DIF: I   OBJ: 2-2.1

Copyright © by Holt, Rinehart and Winston. All rights reserved.

6. How does the cooling rate of lava or magma affect the texture of an igneous rock?

ANS:
When magma cools slowly, crystals have a long time to grow, so the igneous rock that forms is coarse-grained. When magma cools quickly, crystals have a short time to grow, so the igneous rock that forms is fine-grained.

DIF: I   OBJ: 2-2.2

7. Describe the process by which clastic sedimentary rock forms.

ANS:
Clastic sedimentary rock forms when sediments become compacted and cemented together.

DIF: I   OBJ: 2-3.1

8. List three sedimentary rock structures, and explain how they record geologic processes.

ANS:
Three sedimentary rock structures are strata, ripple patterns, and fossils. Strata form when layers of sediment are deposited on top of each other. Ripple patterns form when sediments are shaped by flowing water before they turn into rock. Fossils form from the remains of organisms in sedimentary rock.

DIF: I   OBJ: 2-3.2

9. Both clastic and chemical sedimentary rocks are classified according to texture and composition. Which property is more important for each sedimentary rock type? Explain.

ANS:
Texture is more important in classifying clastic sedimentary rock because clastic sedimentary rock is made of different sizes of sediments. Composition is more important in classifying chemical sedimentary rock because chemical sedimentary rock forms from different materials that crystallize out of solution.

DIF: II   OBJ: 2-3.1

10. What environmental factors cause rock to undergo metamorphism?

ANS:
Increased pressure and increased temperature can cause metamorphism.

DIF: I   OBJ: 2-4.1

Holt Science and Technology
Copyright © by Holt, Rinehart and Winston. All rights reserved.
36

11. What is the difference between foliated and nonfoliated metamorphic rock?

ANS:
The two types of rock differ in texture. Foliated metamorphic rock consists of minerals that are aligned and look like pages in a book. Nonfoliated metamorphic rock does not appear to have any regular patterns.

DIF: I   OBJ: 2-4.3

12. If you had two metamorphic rocks, one with garnet crystals and the other with chlorite crystals, which one would have formed at a deeper level in the Earth's crust? Explain.

ANS:
The rock with garnet crystals would have formed deeper in the Earth because the mineral garnet forms at a higher temperature and a higher pressure than the mineral chlorite.

DIF: II   OBJ: 2-4.2

13. Name four processes that change rock from one type to another.

ANS:
Four processes that change rock from one type to another are weathering, changes in pressure, melting, and cooling.

DIF: I   OBJ: 2-1.2

14. Explain how making glass is similar to the formation of igneous rock.

ANS:
Glass is made by melting quartz sand grains.

DIF: II   OBJ: 2-2.1

15. Explain how mixing concrete and allowing it to harden is similar to the formation of sedimentary rock.

ANS:
Concrete is a mixture of different compounds and rock particles. After it is poured, it hardens much like sedimentary rock.

DIF: II   OBJ: 2-3.1

16. Explain how baking bricks in a kiln is similar to the formation of metamorphic rock.

ANS:
Bricks are made from clay and are baked to make them strong and resistant to weathering. Bricks are "metamorphosed" because they have different properties than dried clay.

DIF: II   OBJ: 2-4.1

Holt Science and Technology
Copyright © by Holt, Rinehart and Winston. All rights reserved.

17. Describe felsic and mafic rocks, and name three elements that occur in each type.

    ANS:
    Felsic rock is lighter in color and weight; it is rich in aluminum, silicon, sodium, and potassium. Mafic rock is darker and heavier; it is rich in iron, magnesium, and calcium.

    DIF: I   OBJ: 2-2.2

18. What's the difference between intrusive and extrusive rock?

    ANS:
    Intrusive rock forms from magma that solidifies while still underground, while extrusive rock forms from magma that solidifies after it has reached the surface.

    DIF: I   OBJ: 2-2.3

19. How does chemical limestone form?

    ANS:
    It forms when calcium carbonate crystallizes out of sea water.

    DIF: I   OBJ: 2-3.1

20. What is stratification, and why is it important to Earth scientists?

    ANS:
    Stratification is the layering of rock. It is important because it records many events in Earth's history as well as erosion and deposition rates.

    DIF: I   OBJ: 2-3.2

21. Explain what a regional metamorphic rock is.

    ANS:
    A regional metamorphic rock has been changed by intense pressure and heat across great regions of the crust rather than by direct contact with magma.

    DIF: I   OBJ: 2-4.1

22. What does the composition of a metamorphic rock tell you about the rock's origin and formation?

    ANS:
    Different metamorphic minerals indicate different temperature and pressure conditions that existed when they formed.

    DIF: I   OBJ: 2-4.2

Copyright © by Holt, Rinehart and Winston. All rights reserved.

23. In no more than three sentences, explain the rock cycle.

ANS:
Answers will vary. Sample answer: In the rock cycle, igneous, sedimentary, and metamorphic rock can change into other rock types through a variety of natural processes. Although rock material changes form, it still remains a part of the rock cycle.

DIF: I   OBJ: 2-1.2

24. How are sandstone and siltstone different from one another? How are they the same?

ANS:
Sandstone has a coarser-grained texture than siltstone. Both sandstone and siltstone are clastic sedimentary rocks.

DIF: I   OBJ: 2-3.1

25. In one or two sentences, explain how the cooling rate of magma affects the texture of the igneous rock that forms.

ANS:
When magma cools slowly, crystals have a long time to grow, so they grow to a much larger size than they do when magma cools quickly.

DIF: I   OBJ: 2-2.1

26. On a separate sheet of paper, create a concept map using the following terms: *rocks, clastic, metamorphic, nonfoliated, igneous, intrusive, chemical, foliated, extrusive, organic, sedimentary.*

ANS:

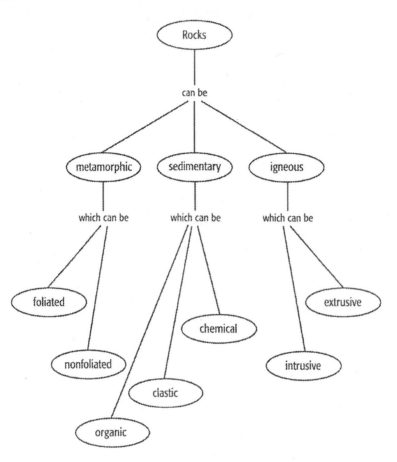

DIF: II   OBJ: 2-4.3

27. The sedimentary rock coquina is made up of pieces of seashells. Which of the three kinds of sedimentary rock could it be? Explain.

ANS:
The seashells that make up coquina are made by shellfish, so coquina is an organic sedimentary rock. (Coquina could also be considered a clastic sedimentary rock because the shell fragments are clasts that have been deposited.)

DIF: II   OBJ: 2-3.1

Holt Science and Technology
Copyright © by Holt, Rinehart and Winston. All rights reserved.

28. If you were looking for fossils in the rocks around your home and the rock type that was closest to your home was metamorphic, would you find many fossils? Why or why not?

ANS:
You would not find many fossils where you lived because fossils are found in sedimentary rock, not metamorphic rock. (Occasionally, fossils are preserved in metamorphic rock that was once sedimentary rock.)

DIF: II   OBJ: 2-4.2

29. Suppose you are writing a book about another planet. In your book, you mention that the planet has no atmosphere or weather. Which type of rock will you not find on the planet? Explain.

ANS:
You will not find sedimentary rock because no weathering of rock can occur because there is no atmosphere.

DIF: II   OBJ: 2-3.1

30. Imagine that you want to quarry or mine granite. You have all of the equipment, but you need a place to quarry. You have two pieces of land to choose from. One piece is described as having a granite batholith under it, and the other has a granite sill. If both plutonic bodies were at the same depth, which one would be a better buy for you? Explain your answer.

ANS:
The property with the batholith would be a better buy because batholiths are much bigger than sills.

DIF: II   OBJ: 2-2.3

31. If a 60 kg granite boulder were broken down into sand grains and if quartz made up 35 percent of the boulder's mass, how many kilograms of the resulting sand would be quartz grains?

ANS:
35% of 60 kg = $60 \times 0.35 = 21$ kg

DIF: II   OBJ: 2-1.3

Holt Science and Technology
Copyright © by Holt, Rinehart and Winston. All rights reserved.

The curve on the graph below shows how the melting point of a particular rock changes with increasing temperature and pressure. Use the graph to answer the questions that follow.

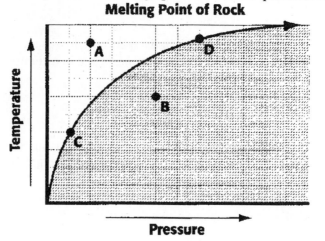

Melting Point of Rock

32. What type of material, liquid or solid, would you find at point A? Why?

   ANS:
   The material at point A is magma. It is magma because everything above the curve on the graph is liquid.

   DIF: II   OBJ: 2-2.2

33. What would you find at point B?

   ANS:
   At point B I would find solid rock.

   DIF: II   OBJ: 2-3.1

34. Points C and D represent different temperature and pressure conditions for a single, solid rock. Why does this rock have a higher melting temperature at point D than it does at point C?

   ANS:
   Although at point D the rock is at a higher temperature, it has much more pressure on it, which keeps it solid.

   DIF: II   OBJ: 2-4.1

35. Describe the basic differences between the formation of metamorphic and igneous rocks.

   ANS:
   Igneous rock forms from molten rock. It solidifies underground or on the Earth's surface. Metamorphic rock forms from rock that is under pressure and heated, but not melted.

   DIF: I   OBJ: 2-4.1

Holt Science and Technology
Copyright © by Holt, Rinehart and Winston. All rights reserved.

36. Describe how two types of sedimentary rock are made.

ANS:
Students should give two of the following: Clastic rock is made of fragments of other rocks and minerals. Chemical sedimentary rock forms when minerals crystallize out of a solution such as sea water. Organic sedimentary rock forms from the remains of organisms.

DIF: I   OBJ: 2-3.1

37. The table shows the percentage by weight of the common elements found in the Earth's crust. In addition to the common elements in the table below, the crust contains other naturally occurring elements. What percentage of the crust's total weight is made up of the other naturally occurring elements? Show your work.

The Most Common Elements in the Earth's Crust

Element	Percentage by weight
Oxygen	46.00
Silicon	27.72
Aluminum	8.13
Iron	5.00
Calcium	3.63
Sodium	2.83
Potassium	2.59
Magnesium	2.09

ANS:
Chart total: 97.99% 100% − 97.99% = 2.01% About two percent of the crust's total weight is made of the remaining elements.

DIF: II   OBJ: 2-1.2

Main Minerals in Some Igneous Rocks

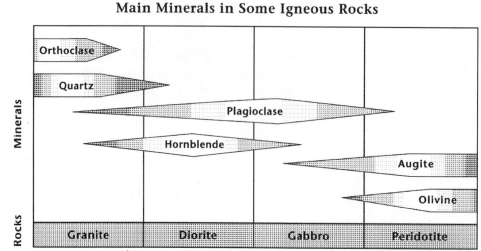

38. Use the information in the chart to compare the compositions of granite and diorite.

    ANS:
    Granite has more quartz and much less plagioclase and hornblende than diorite has. Unlike granite, diorite contains no orthoclase.

    DIF: II   OBJ: 2-2.3

39. Why might sedimentary rock be more common than igneous rock and metamorphic rock on the Earth's surface?

    ANS:
    Sample answer: Sedimentary rock is easily formed from the other two types of rock. When igneous or metamorphic rock erodes, the eroded material can settle and form sedimentary rock. Because metamorphic rock and igneous rock require special conditions to form, they are not as common as sedimentary rock, which can form under a variety of conditions.

    DIF: II   OBJ: 2-1.2

Copyright © by Holt, Rinehart and Winston. All rights reserved.

40. Use the following terms to complete the concept map below: *conglomerate, igneous rock, dikes, texture, erode, sedimentary rock, composition.*

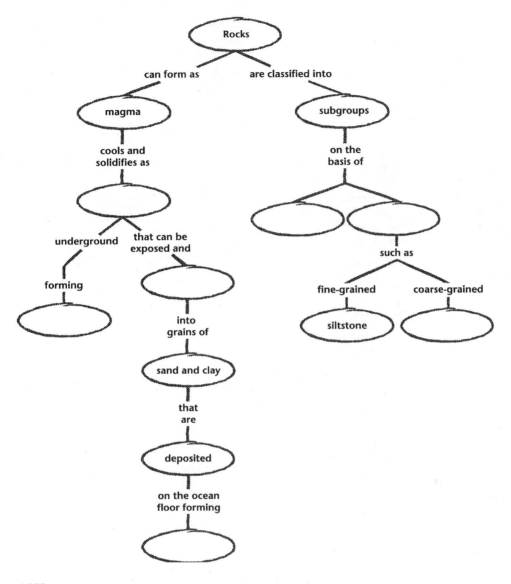

ANS:

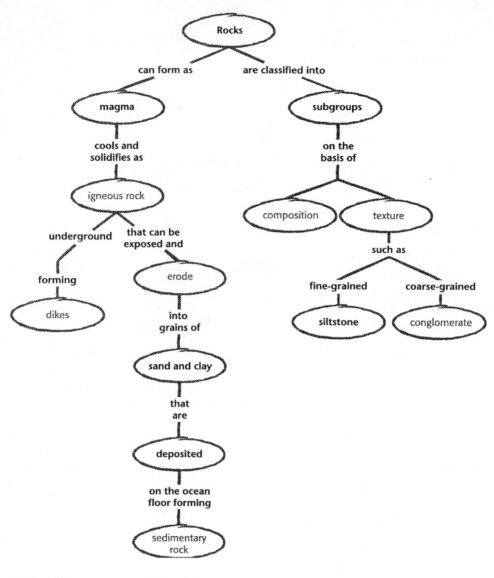

DIF: II   OBJ: 2-1.2

Holt Science and Technology
Copyright © by Holt, Rinehart, and Winston. All rights reserved.

MULTIPLE CHOICE

1. Which of the following words does NOT describe catastrophic geologic change?
   a. sudden
   b. widespread
   c. gradual
   d. time

   ANS: C   DIF: I   OBJ: 3-1.2

2. Scientists assign relative ages by using
   a. potassium-argon dating.
   b. the principle of superposition.
   c. radioactive half-lives.
   d. the ratios of isotopes.

   ANS: B   DIF: I   OBJ: 3-2.1

3. Rock layers cut by a fault formed
   a. after the fault.
   b. before the fault.
   c. at the same time as the fault.
   d. Cannot be determined

   ANS: B   DIF: I   OBJ: 3-2.4

4. If the half-life of an unstable element is 5,000 years, what percentage of the parent material will be left after 10,000 years?
   a. 100
   b. 75
   c. 50
   d. 25

   ANS: D   DIF: II   OBJ: 3-3.2

5. Of the following unstable isotopes, which has the longest half-life?
   a. uranium-238
   b. potassium-40
   c. carbon-14

   ANS: A   DIF: I   OBJ: 3-3.3

6. Fossils can be
   a. petrified.
   b. dried out.
   c. frozen.
   d. All of the above

   ANS: D   DIF: I   OBJ: 3-4.1

7. Of the following geologic time intervals, which is the shortest?
   a. an eon
   b. a period
   c. an era
   d. an epoch

   ANS: D   DIF: I   OBJ: 3-5.1

8. If Earth's history is put on a scale of 12 hours, human civilizations would have been around for
   a. hours.
   b. minutes.
   c. less than 1 second.

   ANS: C   DIF: I   OBJ: 3-5.2

9. Geologists would use ____ to determine the most accurate and precise age of Earth's oldest rocks.
   a. relative dating
   b. the uranium-lead method
   c. the carbon-14 method
   d. index fossils

   ANS: C   DIF: I   OBJ: 3-3.4

10. Which of these is a trace fossil?
    a. a mark left by a dinosaur's tail
    b. a mosquito trapped in amber
    c. a mummified plant seed
    d. a frozen woolly mammoth

    ANS: A   DIF: I   OBJ: 3-4.4

11. According to geologists, the Earth is approximately
    a. 5,000 years old.
    b. 1,000 years old.
    c. 2.5 million years old.
    d. 4.6 billion years old.

    ANS: D   DIF: I   OBJ: 3-5.1

12. Which one of the following processes always occurs at a constant rate?
    a. erosion
    b. decay of organic matter
    c. sediment deposition
    d. radioactive decay

    ANS: D   DIF: I   OBJ: 3-1.3

13. The phrase "younger over older" can be used to remember the principle of
    a. absolute dating.
    b. geologic columns.
    c. unconformities.
    d. superposition.

    ANS: D   DIF: I   OBJ: 3-2.2

14. WHich of the following statements describes geological change?
    a. Geological change is uniform.
    b. Geological change is gradual.
    c. Geological change is sudden.
    d. all of the above

    ANS: D   DIF: I   OBJ: 3-1.1

15. The principle of ____ states that the same geologic processes shaping the Earth today have been at work throughout Earth's history.
    a. uniformitarianism
    b. catastrophism
    c. gradualism
    d. None of the above

    ANS: A   DIF: I   OBJ: 3-1.1

16. The principle that states that all geologic change occurs suddenly is called
    a. gradualism.
    b. catastrophism.
    c. uniformitarianism.
    d. None of the above

    ANS: B   DIF: I   OBJ: 3-1.2

17. A major difference between uniformitarianism and catastrophism is that they predicted
    a. where it is best to dig for fossils.
    b. whether civilizations could die out.
    c. significantly different ages of the Earth.
    d. whether a species could become extinct.

    ANS: C   DIF: I   OBJ: 3-1.2

18. Which theory does modern geology embrace?
    a. uniformitarianism   c. gradualism
    b. catastrophism   d. Both (a) and (b)

    ANS: D   DIF: I   OBJ: 3-1.3

19. Determining whether an object or event is older or younger than other objects or events is called
    a. absolute dating.   c. relative dating.
    b. radiometric dating.   d. None of the above

    ANS: C   DIF: I   OBJ: 3-2.1

20. Suppose you organize photos for an album covering the time period from your birth until your second birthday. No dates have been written on the backs of the pictures, but you are able to put your photos in chronological order by looking at changes in your physical appearance. This is an example of
    a. absolute dating.   c. relative dating.
    b. radiometric dating.   d. None of the above

    ANS: C   DIF: II   OBJ: 3-2.1

21. Suppose you organize your photos for a birthday scrapbook. Although the dates have not been written on the backs of the pictures, each year your family took a photo of you blowing out the candles on your cake. By counting the number of candles on the cake, you are using ____ to organize your photos.
    a. absolute dating   c. relative dating
    b. radiometric dating   d. None of the above

    ANS: A   DIF: II   OBJ: 3-2.1

22. Which activity is an example of the superposition principle?
    a. hanging clothes in a closet   c. choosing a book from the library
    b. tossing clothes into a laundry hamper   d. all of the above

    ANS: B   DIF: II   OBJ: 3-2.2

Copyright © by Holt, Rinehart and Winston. All rights reserved.

Examine the illustration of rock layers below, and answer the questions that follow.

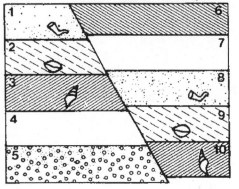

23. Which process disturbed these rock layers?
    a. tilting
    b. folding
    c. faulting
    d. intrusion

    ANS: C   DIF: II   OBJ: 3-2.4

24. Which rock layers have the same relative age?
    a. **1 and 6**
    b. **2 and 9**
    c. **3 and 7**
    d. **5 and 10**

    ANS: B   DIF: II   OBJ: 3-2.4

25. Which of the following would most likely be used to determine the relative age of each rock layer?
    a. the geologic column
    b. superposition
    c. faults
    d. all of the above

    ANS: D   DIF: II   OBJ: 3-2.4

26. Missing rock layers that create gaps in rock-layer sequences are called
    a. intrusions.
    b. uniformities.
    c. unconformities.
    d. folds.

    ANS: C   DIF: I   OBJ: 3-2.4

27. Which process might create unconformities in rock layers?
    a. nondeposition of sediments
    b. erosion of sediments
    c. deposition of sediments
    d. both (a) and (b)

    ANS: D   DIF: II   OBJ: 3-2.4

28. The most common type of unconformity is called a(n) _____
    a. disconformity.
    b. nonconformity.
    c. angular unconformity.
    d. None of the above

    ANS: A   DIF: I   OBJ: 3-2.4

Holt Science and Technology
Copyright © by Holt, Rinehart and Winston. All rights reserved.

29. A(n) ____ exists where sedimentary rock layers lie on top of an eroded surface of non-layered igneous or metamorphic rock.
   a. angular unconformity
   b. nonconformity
   c. disconformity
   d. linear unconformity

   ANS: B   DIF: I   OBJ: 3-2.4

30. Which of the following is a type of unconformity?
   a. disconformity
   b. nonconformity
   c. angular unconformity
   d. all of the above

   ANS: D   DIF: I   OBJ: 3-2.4

31. A(n) ____ exists between horizontal rock layers and rock layers that are tilted or folded.
   a. disconformity
   b. nonconformity
   c. angular unconformity
   d. linear unconformity

   ANS: C   DIF: I   OBJ: 3-2.4

Examine the diagram of rock layers below, and answer the questions that follow.

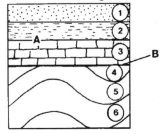

32. To determine the relative age of rock layers **1–3**, you could use the principle of
   a. uniformitarianism.
   b. superposition.
   c. catastrophism.
   d. None of the above

   ANS: B   DIF: II   OBJ: 3-2.2

33. Rock layers **4–6** in have been disturbed by
   a. faulting
   b. intrusion.
   c. folding.
   d. tilting.

   ANS: C   DIF: II   OBJ: 3-2.4

34. The boundary labeled **A** represents a type of unconformity that is often hard to see called
   a. a disconformity.
   b. a nonconformity.
   c. an angular unconformity.
   d. None of the above

   ANS: A   DIF: II   OBJ: 3-2.4

35. Which type of unconformity does the boundary labeled **B** represent?
   a. disconformity
   b. nonconformity
   c. angular unconformity
   d. all of the above

   ANS: C   DIF: II   OBJ: 3-2.4

36. What could you conclude if layer **3** contains reptile fossils and layer **5** contains amphibian fossils?
    a. Reptiles existed before amphibians.
    b. Amphibians existed before reptiles.
    c. Reptiles existed at the same time as amphibians.
    d. none of the above

    ANS: B   DIF: II   OBJ: 3-2.4

37. ___ is a process of establishing the age of an object, such as a fossil or rock layer, by determining the number of years it has existed.
    a. Absolute dating   c. Relative dating
    b. Radiometric dating   d. All of the above

    ANS: A   DIF: I   OBJ: 3-3.2

38. Determining the absolute age of a sample based on the ratio of parent material to daughter material is called
    a. hypothetical dating.   c. relative dating.
    b. radiometric dating.   d. None of the above

    ANS: B   DIF: I   OBJ: 3-3.2

39. One-fourth of an ivory carving's original carbon-14 remains unchanged. Carbon-14 has a half-life of 5,730 years. How old is the carving?
    a. 1,432.5 years   c. 5,730 years
    b. 4,297.5 years   d. 11,460 years

    ANS: D   DIF: II   OBJ: 3-3.2

40. A bowl carved out of bone is displayed in a museum. Half of its original carbon-14 has decayed. Carbon-14 has a half-life of 5,730 years. How old is the bowl?
    a. 1,432.5 years   c. 5,730 years
    b. 4,297.5 years   d. 11,460 years

    ANS: C   DIF: II   OBJ: 3-3.2

41. A human bone has been recently excavated from an archaeological dig. One-eighth of the bone's original carbon-14 remains. Carbon-14 has a half-life of 5,730 years. How old is the bone?
    a. 716 years   c. 5,730 years
    b. 2,865 years   d. 17,190 years

    ANS: D   DIF: II   OBJ: 3-3.2

42. A mummy was recently discovered in a catacomb. Suppose half of the original carbon-14 atoms in the mummy remain. Carbon-14 has a half-life of 5,730 years. How old is the mummy?
    a. 5,730 years   c. 1,432.5 years
    b. 2,865 years   d. 17,190 years

    ANS: A   DIF: II   OBJ: 3-3.2

43. Which type of radiometric dating is used to date rocks older than 10 million years?
    a. carbon-14 method
    b. potassium-argon method
    c. uranium-lead method
    d. all of the above

    ANS: C   DIF: I   OBJ: 3-3.4

44. Which type of radiometric dating would be best to use to date rocks older than 100,000 years?
    a. carbon-14 method
    b. potassium-argon method
    c. uranium-lead method
    d. all of the above

    ANS: B   DIF: I   OBJ: 3-3.4

45. Which type of radiometric dating would be best to use to date rocks younger than 50,000 years?
    a. carbon-14 method
    b. potassium-argon method
    c. uranium-lead method
    d. all of the above

    ANS: A   DIF: I   OBJ: 3-3.4

46. Which of the following is a radioactive isotope of carbon?
    a. carbon-14
    b. carbon-13
    c. carbon-12
    d. all of the above

    ANS: A   DIF: I   OBJ: 3-3.3

47. The decay product of uranium-238 is
    a. potassium-40.
    b. lead-206.
    c. carbon-14.
    d. carbon-12.

    ANS: B   DIF: III   OBJ: 3-3.3

48. Which of the following is NOT a medium in which organisms fossilize?
    a. amber
    b. a glacier
    c. ocean water
    d. tar

    ANS: C   DIF: I   OBJ: 3-4.1

49. Which of the following organisms would be most likely to fossilize when buried in sediment?
    a. an organism with a soft body
    b. an organism with a hard shell
    c. an organism with weblike wing structures
    d. a unicellular organism, such as an amoeba

    ANS: B   DIF: I   OBJ: 3-4-1

50. Organisms can be preserved by _____, a process in which minerals fill in pore spaces of an organism's tissues.
    a. mummification
    b. petrification
    c. permineralization
    d. freezing

    ANS: C   DIF: I   OBJ: 3-4.1

51. _____ of an organism occurs when its tissues are completely replaced by minerals.
    a. Mummification
    b. Petrification
    c. Permineralization
    d. Freezing

    ANS: B   DIF: I   OBJ: 3-4.1

52. Whichof the following is the process by which trees fossilize?
    a. mummification
    b. petrification
    c. permineralization
    d. freezing

    ANS: B   DIF: I   OBJ: 3-4.1

53. Insects can be preserved in _____, or hardened tree sap.
    a. amber
    b. tar
    c. ice
    d. coprolite

    ANS: A   DIF: I   OBJ: 3-4.1

54. Much of what we know about saber-toothed cats has come from animals that fossilized in
    a. rock.
    b. tar.
    c. ice.
    d. amber.

    ANS: B   DIF: I   OBJ: 3-4.1

55. The remains of organisms that have died in dry places are preserved through a process known as
    a. mummification.
    b. petrification.
    c. permineralization.
    d. freezing.

    ANS: A   DIF: I   OBJ: 3-4.1

56. _____ fossil specimens are some of the best preserved of all fossils.
    a. Petrified
    b. Mummified
    c. Frozen
    d. Permineralized

    ANS: C   DIF: I   OBJ: 3-4.1

57. Which of the following is a type of fossil that was NOT part of an organism?
    a. an insect in amber
    b. a mummified seed
    c. a trace fossil
    d. petrified wood

    ANS: C   DIF: I   OBJ: 3-4.2

58. A _____ is a cavity in the ground or rock where a plant or animal was buried.
    a. mold
    b. cast
    c. track
    d. burrow

    ANS: A   DIF: I   OBJ: 3-4.2

Copyright © by Holt, Rinehart and Winston. All rights reserved.

59. A _____ is an object created when sediment fills a mold and becomes rock.
    a. burrow
    b. cast
    c. coprolite
    d. track

    ANS: B   DIF: I   OBJ: 3-4.2

60. A _____ remains when an animal's footprint fills with sediment that eventually turns to rock.
    a. burrow
    b. cast
    c. track
    d. mold

    ANS: C   DIF: I   OBJ: 3-4.2

61. A _____ is a fossilized shelter made by an animal that dug into the ground.
    a. coprolite
    b. cast
    c. track
    d. burrow

    ANS: D   DIF: I   OBJ: 3-4.2

62. A _____ provides valuable information about the habits and diets of the animals that left it.
    a. coprolite
    b. cast
    c. track
    d. burrow

    ANS: A   DIF: I   OBJ: 3-4.2

63. Scientists often find rocks that contain marine fossils on mountaintops. What could such a discovery allow a scientist to conclude about how the position of the rocks changed over time?
    a. The position of the rock layers has not changed over time.
    b. Sediments were eroded from the ocean floor and redeposited here.
    c. These rocks were once below the surface of an ocean and were pushed up from below sea level.
    d. These rocks were once buried under a glacier.

    ANS: C   DIF: II   OBJ: 3-4.3

64. The largest divisions of geologic time are called
    a. epochs.
    b. eons.
    c. eras.
    d. periods.

    ANS: B   DIF: I   OBJ: 3-5.1

65. The four eons are divided into
    a. epochs.
    b. periods.
    c. eras.
    d. years.

    ANS: C   DIF: I   OBJ: 3-5.1

66. Eras are divided into
    a. eons.
    b. periods.
    c. years.
    d. epochs.

    ANS: B   DIF: I   OBJ: 3-5.1

67. Some periods are divided into
    a. epochs.
    c. years.
    b. eras.
    d. eons.

    ANS: A   DIF: I   OBJ: 3-5.1

68. The ____ is characterized by the first land-dwellers—plants and amphibians.
    a. Cenozoic era
    c. Paleozoic era
    b. Mesozoic era
    d. Hadean eon

    ANS: C   DIF: I   OBJ: 3-5.2

69. The ____ is known as the Age of Reptiles.
    a. Cenozoic era
    c. Paleozoic era
    b. Mesozoic era
    d. Hadean eon

    ANS: B   DIF: I   OBJ: 3-5.2

70. The ____ is known as the Age of Mammals.
    a. Cenozoic era
    c. Paleozoic era
    b. Mesozoic era
    d. Hadean eon

    ANS: A   DIF: I   OBJ: 3-5.2

71. In which eon do we live?
    a. Phanerozoic eon
    c. Archean eon
    b. Proterozoic eon
    d. Hadean eon

    ANS: A   DIF: I   OBJ: 3-5.3

72. In which eon did the first organisms with well-developed cells appear?
    a. Phanerozoic eon
    c. Archean eon
    b. Proterozoic eon
    d. Hadean eon

    ANS: B   DIF: I   OBJ: 3-5.2

73. In which eon did the earliest known rocks on Earth form?
    a. Phanerozoic eon
    c. Archean eon
    b. Proterozoic eon
    d. Hadean eon

    ANS: C   DIF: I   OBJ: 3-5.2

74. The only rocks that scientists have found from the ____ are meteorites and rocks from the moon.
    a. Phanerozoic eon
    c. Archean eon
    b. Proterozoic eon
    d. Hadean eon

    ANS: D   DIF: I   OBJ: 3-5.2

Copyright © by Holt, Rinehart and Winston. All rights reserved.

COMPLETION

1. According to the principle of _____, geologic processes operate today as they did in Earth's past. (uniformitarianism or catastrophism)

   ANS: uniformitarianism   DIF: I   OBJ: 3-1.1

2. A trilobite was buried by ocean sediment, leaving a cavity, or _____. (cast or mold)

   ANS: mold   DIF: I   OBJ: 3-4.2

3. The process of _____ involves the comparison of one rock layer with others in a sequence. (relative dating or absolute dating)

   ANS: relative dating   DIF: I   OBJ: 3-2.1

4. The largest divisions of geologic time are called _____. (eons or eras)

   ANS: eons   DIF: I   OBJ: 3-5.1

5. Erosion is a major cause of the missing rock layers known as _____. (superposition or unconformities)

   ANS: unconformities   DIF: I   OBJ: 3-2.5

6. _____ is the branch of science that studies the history of the Earth.

   ANS: Geology   DIF: I   OBJ: 3-1.1

7. _____ is a principle that states that younger rocks lie above older rocks in undisturbed sequences.

   ANS: Superposition   DIF: I   OBJ: 3-2.2

8. The _____ is an ideal sequence of rock layers that contains all the known fossils and rock formations on Earth, arranged from oldest to youngest.

   ANS: geologic column   DIF: I   OBJ: 3-2.3

9. A _____ is a break in the Earth's crust along which blocks of the crust slide relative to one another.

   ANS: fault   DIF: I   OBJ: 3-2.4

Copyright © by Holt, Rinehart and Winston. All rights reserved.

10. An _____ is molten rock from the Earth's interior that squeezes into existing rock and cools.

ANS: intrusion   DIF: I   OBJ: 3-2.4

11. _____ occurs when rock layers bend and buckle from Earth's internal forces.

ANS: Folding   DIF: I   OBJ: 3-2.4

12. _____ occurs when internal forces in the Earth slant rock layers without folding them.

ANS: Tilting   DIF: I   OBJ: 3-2.4

13. _____ are atoms of the same element that have the same number of protons but have different numbers of neutrons.

ANS: Isotopes   DIF: I   OBJ: 3-3.1

14. The process of _____ occurs when radioactive isotopes break down into stable isotopes of other elements.

ANS: radioactive decay   DIF: I   OBJ: 3-3.1

15. The time it takes for one-half of a radioactive sample to decay is called a(n) _____.

ANS: half-life   DIF: I   OBJ: 3-3.2

16. A(n) _____ is any naturally preserved evidence of life.

ANS: fossil   DIF: I   OBJ: 3-4.1

17. Any naturally preserved evidence of an animal's activity is called a(n) _____.

ANS: trace fossil   DIF: I   OBJ: 3-4.2

18. Preserved feces, or dung, from animals are called _____.

ANS: coprolites   DIF: I   OBJ: 3-4.2

19. _____ are fossils of organisms that lived during a relatively short, well-defined time span.

ANS: Index fossils   DIF: I   OBJ: 3-4.4

20. The _____ is a scale that divides Earth's 4.6-billion-year history into distinct intervals of time.

ANS: geologic time scale   DIF: I   OBJ: 3-5.1

SHORT ANSWER

For each pair of terms, explain the difference in their meaning.

1. geologic time scale/geologic column

ANS:
The geologic time scale is the history of the Earth divided into eons, eras, periods, and epochs. The geologic column is an idealized sequence of rock layers than contains all known fossils and rock formations arranged from oldest to youngest.

DIF: I   OBJ: 3-5.1

2. eon/era

ANS:
An eon is the largest division of geologic time. The Phanerozoic eon is divided into three eras.

DIF: I   OBJ: 3-5.1

3. mold/cast

ANS:
A mold is the cavity created in ground or rock where an organism was buried. A cast forms when material fills the mold and becomes rock.

DIF: I   OBJ: 3-4.2

4. relative dating/absolute dating

ANS:
Relative dating is a method of comparing the age of a rock or fossil with the age of other objects or events. For example, a fault is always younger than the layers it disturbs. Absolute dating is a method of determining the age of something in years.

DIF: I   OBJ: 3-3.1

5. uniformitarianism/catastrophism

ANS:
Uniformitarianism is the theory that gradual geologic processes that we observe in the present were also active in the past. This theory argues that slow gradual change shapes the Earth. Catastrophism is the theory that past episodes of sudden and drastic change are responsible for the major geologic features of the Earth.

DIF: I   OBJ: 3-1.2

6. Why do Earth scientists need the principle of uniformitarianism in order to make predictions?

ANS:
Scientists make predictions based on the past as well as the present. To make predictions, they must assume that geologic processes will be similar in the future.

DIF: I   OBJ: 3-1.1

7. What is the difference between uniformitarianism and catastrophism?

ANS:
Catastrophism states that the geologic history of the Earth was dominated by sudden, drastic changes that built features such as mountains, valleys, and oceans. Uniformitarianism argues that the Earth is shaped by slow, gradual processes that can still be observed today.

DIF: I   OBJ: 3-1.2

8. How was the role of catastrophism in Earth science changed?

ANS:
Geologists now agree that sudden catastrophic events such as asteroid impacts or volcanic eruptions can also cause geologic change.

DIF: II   OBJ: 3-1.3

9.
    a. In a rock-layer sequence that has not been disturbed, are older layers found on top of younger layers?
    b. What rule did you use to answer a?

ANS:
a. No; the younger layers are found on top.
b. I used the principle of superposition to answer a.

DIF: I   OBJ: 3-2.2

Holt Science and Technology
Copyright © by Holt, Rinehart and Winston. All rights reserved.

10. List five events or features that can disturb rock-layer sequences.

ANS:
Answers will vary but should include a fault, an intrusion, a disconformity, a nonconformity, and an angular unconformity.

DIF: I   OBJ: 3-2.4

11. Consider a fault that cuts through all the layers of a rock-layer sequence. Is the fault older or younger than the layers? Explain.

ANS:
The fault is younger than the layers. The layers had to be present for the fault to cut across them.

DIF: I   OBJ: 3-2.5

12. Unlike other types of unconformities, disconformities are hard to recognize because all the layers are horizontal. How does a geologist know when he or she is looking at a disconformity?

ANS:
Disconformities represent a gap in the geologic column. If part of the column is missing from the layers, then the geologist has observed a disconformity.

DIF: II   OBJ: 3-2.5

13. Explain how radioactive decay occurs.

ANS:
Radioactive decay occurs as a radioactive isotope breaks down into a stable isotope. This happens as the isotope loses an electron and a neutron becomes a proton.

DIF: I   OBJ: 3-3.1

14. How does radioactive decay relate to radiometric dating?

ANS:
Radioactive decay occurs at a constant rate. By determining the ratio between the parent material and the daughter material in an object, scientists can determine how old the object is.

DIF: I   OBJ: 3-3.2

15. List three types of radiometric dating.

ANS:
Three types of radiometric dating are uranium-lead, carbon-14, and potassium-argon.

DIF: I   OBJ: 3-3.3

Copyright © by Holt, Rinehart and Winston. All rights reserved.

16. Which radiometric-dating method would be most appropriate for dating artifacts found at Effigy Mounds? Explain.

ANS:
Carbon-14 would be the best method to date artifacts from Effigy Mounds. This is because carbon-14 has a relatively short half-life of 5,730 years.

DIF: II   OBJ: 3-3.4

17. Describe two ways that fossils can form.

ANS:
Answers will vary. Fossils can form as organisms are deposited in layers of sediment. If the sediment becomes rock, the organism may be preserved as a fossil. Permineralization is another way fossils can form. In this process, minerals form in the pore spaces between an organism's tissues. Petrification occurs if the tissues are completely replaced by minerals.

DIF: I   OBJ: 3-4.1

18. List two types of fossils that are not part of an organism.

ANS:
Answers will vary. Students may mention animal burrows, coprolites, or animal tracks.

DIF: I   OBJ: 3-4.2

19. a. What are index fossils?
    b. How do scientists use them to date rocks?

ANS:

a. Index fossils are fossils of organisms that lived during a relatively short, defined period of time.
b. Index fossils help scientists date a rock layer without directly using radiometric dating.

DIF: I   OBJ: 3-4.4

20. If you find rock layers containing fish fossils in a desert, what can you infer about that area of the desert?

ANS:
Answers will vary. The desert was once an ocean, stream, or lake.

DIF: II   OBJ: 3-4.3

Copyright © by Holt, Rinehart and Winston. All rights reserved.

21. How many eras are in the Phanerozoic eon? List them.

ANS:
There are three eras: Cenozoic, Mesozoic, and Paleozoic.

DIF: I   OBJ: 3-5.1

22. Extinctions at the end of two geologic time intervals are mentioned in the chapter. What are these two intervals, and when did each interval end?

ANS:
The Paleozoic era ended about 248 million years ago, and the Mesozoic era ended about 65 million years ago.

DIF: I   OBJ: 3-5.2

23. Which eon do we know the most about? Why?

ANS:
We know the most about the Phanerozoic eon because the rock and fossil record from that eon is the most detailed and complete.

DIF: I   OBJ: 3-5.3

24. What future event might mark the end of the Cenozoic era?

ANS:
Answers will vary. Sample answers: the extinction of humans; the extinction of most wild animal species; the evolution of a new, widespread group of organisms; a major change in the Earth's climate

DIF: II   OBJ: 3-5.3

25. What is catastrophism?

ANS:
Catastrophism is the idea that geologic change occurred suddenly as a result of infrequent disastrous events.

DIF: I   OBJ: 3-1.2

26. Describe uniformitarianism.

ANS:
Uniformitarianism is the view that the Earth is shaped by gradual changes that are still occurring today.

DIF: I   OBJ: 3-1.1

27. Explain what an unconformity is. Give an example.

ANS:
Unconformities are gaps in an area's geologic column. Examples include disconformities, nonconformities, and angular unconformities.

DIF: I   OBJ: 3-2.5

28. How do uniformitarianism and the theory of evolution support each other?

ANS:
Answers will vary. Students should note that both theories suggested that the Earth is much older than previously thought. In addition, they both state that change occurs gradually over time.

DIF: II   OBJ: 3-1.1

29. When using the carbon-14 dating method, which sample would be older, a sample with a ratio of carbon-14 to carbon-12 of 2:1, or a sample with a ratio of 3:1?

ANS:
The sample with a 2:1 ratio would be older.

DIF: II   OBJ: 3-3.3

30. What is a half-life?

ANS:
A half-life is the time it takes for one-half of a radioactive isotope to decay.

DIF: I   OBJ: 3-3.1

31. Would a shark tooth make a good index fossil? Why or why not?

ANS:
A shark tooth would not make a good index fossil because sharks have existed for more than 200 million years; index fossils are useful because the organisms existed for a short period of time.

DIF: I   OBJ: 3-4.4

32. Why do the frigid temperatures of Siberia and the sticky tar of the La Brea Tar Pits preserve fossils so well?

ANS:
Both environments retard the decay of an organism and help preserve it.

DIF: I   OBJ: 3-4.1

Copyright © by Holt, Rinehart and Winston. All rights reserved.

33. What are the largest divisions of time in the geologic time scale?

ANS:
Eons are the largest divisions of time.

DIF: I   OBJ: 3-5.1

34. During which era did plants start to appear on land?

ANS:
Plants started to appear on land during the Paleozoic era.

DIF: I   OBJ: 3-5.2

35. Which group of animals has been on Earth for a longer period of time, birds or humans?

ANS:
Birds have been on Earth for a longer period of time.

DIF: I   OBJ: 3-5.2

36. Describe how plant and animal remains become petrified.

ANS:
Petrification occurs as the tissues of organisms are completely replaced by minerals.

DIF: I   OBJ: 3-4.1

37. Explain how a fossil cast forms.

ANS:
A fossil cast forms when sediment fills in a fossil mold and becomes rock.

DIF: I   OBJ: 3-4.4

33. Use the following terms to create a concept map: *age, absolute dating, half-life, radioactive decay, radiometric dating, relative dating, superposition, geologic column, isotopes.*

ANS:

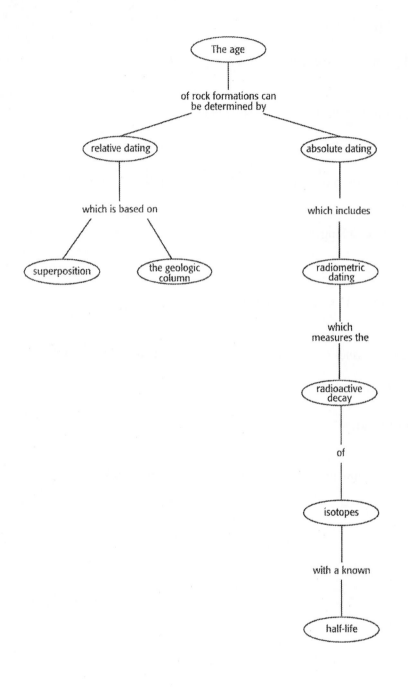

DIF: II   OBJ: 3-3.3

Copyright © by Holt, Rinehart, and Winston. All rights reserved.

39. You may have heard the term *petrified wood.* Why doesn't a "petrified" tree contain any wood?

ANS:
The petrified tree does not contain any wood because the wood tissue in the tree was completely replaced by minerals.

DIF: II   OBJ: 3-3.1

40. How do tracks and burrows end up in the rock and fossil record?

ANS:
Animals leave tracks and create burrows in soil. Sediment fills in these features and buries them quickly. Over time, the sediment becomes rock and preserves these trace fossils.

DIF: II   OBJ: 3-3.2

41. How do you know that an intrusion is younger than its surrounding rock layers?

ANS:
The intrusion is always younger because the rock layers had to be present before the intrusion could disturb them.

DIF: II   OBJ: 3-2.4

42. On the graph below, place a dot on the y-axis at 100 percent. Then place a dot on the graph at each half-life to show how much of the parent material is left. Connect the points with a curved line. Will the percentage of parent material ever reach zero? Explain.

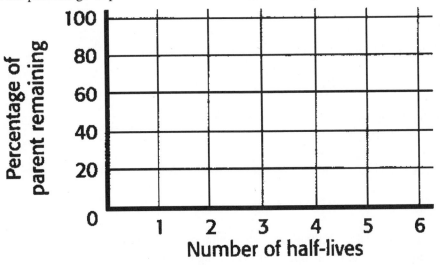

ANS:
Mathematically, the percentage of parent material will never reach zero. In a real sample, however, all of the parent material will eventually decay.

DIF: II   OBJ: 3-3.2

Holt Science and Technology
Copyright © by Holt, Rinehart and Winston. All rights reserved.

Examine the drawing below, and answer the following questions.

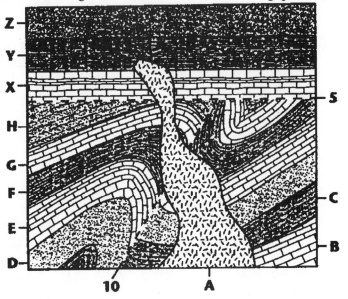

43. Is intrusion A younger or older than layer X?

ANS:
Intrusion A is younger than layer X.

DIF: II   OBJ: 3-2.5

44. What kind of unconformity is marked by 5?

ANS:
5 marks an angular unconformity.

DIF: II   OBJ: 3-2.5

45. Is intrusion A younger or older than fault 10. Why?

ANS:
Intrusion A is younger than fault 10 because the intrusion is not disturbed by the fault.

DIF: II   OBJ: 3-2.5

46. Other than the intrusion and faulting, what event occurred in layers B, C, D, E, F, G, and H?
Number this event, the intrusion, and the faulting in the order they occurred.

ANS:
Erosion; the fault occurred first, then layers eroded, and finally the intrusion eroded.

DIF: II   OBJ: 3-2.5

Holt Science and Technology
Copyright © by Holt, Rinehart and Winston. All rights reserved.

47. "The present is the key to the past." Explain this sentence in terms of understanding Earth's past.

ANS:
According to the principle of uniformitarianism, geological processes operate today as they did in Earth's past. If you understand how these processes work today, you can draw conclusions about how Earth was transformed over time.

DIF: I   OBJ: 3-1.1

48. Explain the principles of superposition. How is this principle useful to geologists?

ANS:
The principle of superposition states that younger rocks lie above older rocks in undisturbed sequences. Geologists use this principle to determine the relative ages of rock layers that have been disturbed by a variety of geological processes.

DIF: I   OBJ: 3-2.2

49. Describe the geologic column and how it is assembled.

ANS:
The geologic column is an ideal rock-layer sequence that contains all the known fossils and rock formations on Earth arranged from oldest to youngest. Geologists create this reference column by merging data from rock sequences collected all over the world.

DIF: I   OBJ: 3-2.3

50. The pictures below show geologic changes. However, the pictures are out of order. Label the pictures with the letters A, B, C, D, and E to show the correct geological sequence. The first picture has been labeled for you. Justify your ordering by explaining the changes that occurred between pictures.

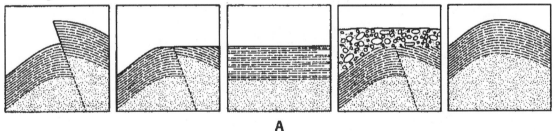

A

ANS:
Pictures are labeled from left to right: C, D, A, E, B. Folding has occurred between pictures A and B. By picture C, the folded layers have broken, and one side has dropped down. Picture D shows erosion of the rock layers exposed by movement along the fault surface. In picture E, sediments have been deposited on top of the eroded surface.

DIF: II   OBJ: 3-2.4

51. In the Paleozoic era, there were dragonflies with 70 cm wingspans. How many feet did the wings of these dragonflies span? Show your work. (Hint: 2.5 cm = 1 in., and 1 ft = 12 in.)

ANS:
70 cm × (1 in. ÷ 2.5 cm) = 28 in.; 28 in. × (1 ft ÷ 12 in.) = 2.3 ft
The dragonfly's wings spanned 2.3 ft.

DIF: II   OBJ: 3-5.2

52. Radon-222 has a half-life of approximately 4 days. If you begin with a sample containing 32 grams of radon, how many grams of radon will be left after 20 days? Show your work.

ANS:
20 days is equivalent to 5 half-lives (20 ÷ 4 = 5). The progression would be
32 g → 16 g → 8 g → 4 g → 2 g → 1 g of radon remaining after 20 days.

DIF: II   OBJ: 3-3.2

Holt Science and Technology
Copyright © by Holt, Rinehart and Winston. All rights reserved.

53. Use the following terms to complete the concept map below: *Phanerozoic, Cenozoic, reptiles, Archean, mammals, eons, Paleozoic.*

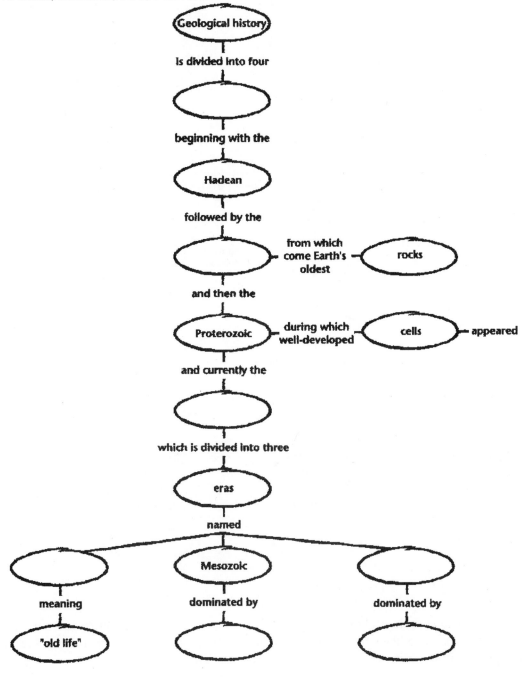

ANS:

Copyright © by Holt, Rinehart and Winston. All rights reserved.

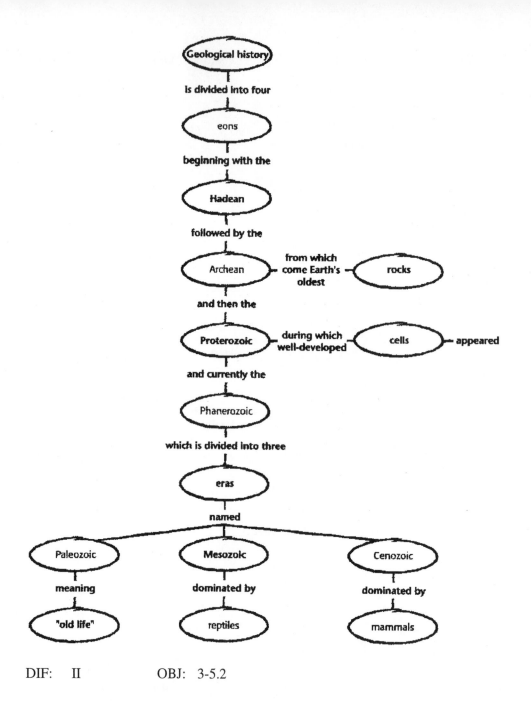

DIF: II   OBJ: 3-5.2

54. Suppose an archaeological investigation uncovers the remains of a campfire. Is the project archaeologist more likely to determine the campfire's age by using carbon-14 dating techniques or potassium-40 dating techniques? Explain your answer.

ANS:
The project archaeologist is more likely to use carbon-14 dating techniques because it has a half-life of 5,730 years and can be used to date objects up to 50,000 years old. Potassium-40 has a half-life of approximately 1.3 billion years and is more useful for dating older objects.

DIF: II   OBJ: 3-3.3

Holt Science and Technology
Copyright © by Holt, Rinehart and Winston. All rights reserved.

TRUE/FALSE

1. The crust is the Earth's only solid layer.

    ANS: F   DIF: I   OBJ: 4-1.2

2. The inner core of the Earth is solid and made primarily of iron.

    ANS: T   DIF: I   OBJ: 4-1.1

3. Temperature and pressure increase toward the center of the Earth.

    ANS: T   DIF: I   OBJ: 4-1.2

4. The asthenosphere is the thinnest layer.

    ANS: F   DIF: I   OBJ: 4-1.2

MULTIPLE CHOICE

1. The part of the Earth that is a liquid is the
    a. crust.   c. outer core.
    b. mantle.   d. inner core.

    ANS: C   DIF: I   OBJ: 4-1.2

2. The part of the Earth on which the tectonic plates are able to move is the
    a. lithosphere.   c. mesosphere.
    b. asthenosphere.   d. subduction zone.

    ANS: B   DIF: I   OBJ: 4-1.2

3. The ancient continent that contained all the landmasses is called
    a. Pangaea.   c. Laurasia.
    b. Gondwana.   d. Panthalassa.

    ANS: A   DIF: I   OBJ: 4-2.1

4. The type of tectonic plate boundary involving a collision between two tectonic plates is
    a. divergent.   c. convergent.
    b. transform.   d. normal.

    ANS: C   DIF: I   OBJ: 4-3.2

5. The type of tectonic plate boundary that sometimes has a subduction zone is
   a. divergent.   c. convergent.
   b. transform.   d. normal.

   ANS: C   DIF: I   OBJ: 4-3.2

6. The San Andreas fault is an example of a
   a. divergent boundary.   c. convergent boundary.
   b. transform boundary.   d. normal boundary.

   ANS: B   DIF: I   OBJ: 4-3.2

7. When a fold is shaped like an arch, with the fold in an upward direction, it is called a(n)
   a. monocline.   c. syncline.
   b. anticline.   d. decline.

   ANS: B   DIF: I   OBJ: 4-4.1

8. The type of fault in which the hanging wall moves down relative to the footwall is called
   a. strike-slip.   c. normal.
   b. reverse.   d. fault block.

   ANS: C   DIF: I   OBJ: 4-4.2

9. The type of mountain involving huge sections of the Earth's crust being pushed up into anticlines and synclines is the
   a. folded mountain.   c. volcanic mountain.
   b. fault-block mountain.   d. strike-slip mountain.

   ANS: A   DIF: I   OBJ: 4-4.3

10. Continental mountain ranges are usually associated with
    a. divergent boundaries.   c. convergent boundaries.
    b. transform boundaries.   d. normal boundaries.

    ANS: C   DIF: I   OBJ: 4-3.2

11. Mid-ocean ridges are associated with
    a. divergent boundaries.   c. convergent boundaries.
    b. transform boundaries.   d. normal boundaries.

    ANS: A   DIF: I   OBJ: 4-3.2

12. Which of the following makes up most of the Earth's mass?
    a. crust   c. outer core
    b. mantle   d. inner core

    ANS: B   DIF: I   OBJ: 4-1.1

Copyright © by Holt, Rinehart and Winston. All rights reserved.

13. Fossils of *Lystrosaurus*, an early land-dwelling reptile, have been found in Antarctica, India, and South Africa. The distribution of these fossils suggests that these areas were once
    a. made of the same chemical elements.
    b. covered by oceanic crust.
    c. home to a wide variety of organisms.
    d. connected to one another.

    ANS: D   DIF: I   OBJ: 4-2.1

14. The speed of seismic waves depends on the _____ of the layer through which they travel.
    a. density   c. shape
    b. thickness   d. position

    ANS: A   DIF: I   OBJ: 4-1.4

15. Magma that reaches the Earth's surface may form _____ mountains.
    a. fault-block   c. volcanic
    b. all   d. subducted

    ANS: C   DIF: I   OBJ: 4-4.4

16. Which layer of the Earth is made up of tectonic plates?
    a. core   c. asthenosphere
    b. mesosphere   d. lithosphere

    ANS: D   DIF: I   OBJ: 4-1.3

17. Which of the following appears to cause movement of Earth's tectonic plates?
    a. convection currents below the lithosphere
    b. energy from volcanic activity
    c. magnetic-pole reversals
    d. faults in mountain ranges

    ANS: A   DIF: I   OBJ: 4-3.1

18. Earth's lightest materials make up the
    a. crust.   c. mantle.
    b. inner core.   d. outer core.

    ANS: A   DIF: I   OBJ: 4-1.1

19. Continental crust has an average thickness of
    a. 5 to 8 km.   c. 30 km.
    b. 5 to 100 km.   d. None of the above

    ANS: C   DIF: II   OBJ: 4-1.1

20. Earth's oceanic crust is _____ than the continental crust.
    a. thinner but denser   c. thicker but less dense
    b. thicker and denser   d. thinner and less dense

    ANS: A   DIF: II   OBJ: 4-1.1

Copyright © by Holt, Rinehart and Winston. All rights reserved.

21. Scientists have learned that the mantle's composition has large amounts of
    a. oxygen and nitrogen.
    b. iron and magnesium.
    c. water.
    d. iron and oxygen.

    ANS: B   DIF: I   OBJ: 4-1.1

22. Scientists believe that the Earth's core is made mostly of
    a. iron.
    b. magnesium.
    c. aluminum.
    d. oxygen.

    ANS: A   DIF: I   OBJ: 4-1.1

23. The diameter of the Earth's core is slightly larger than the diameter of
    a. Saturn.
    b. Jupiter.
    c. Mars.
    d. Uranus.

    ANS: C   DIF: I   OBJ: 4-1.1

24. The outermost, rigid layer of the Earth is called the
    a. mesosphere.
    b. lithosphere.
    c. asthenosphere.
    d. outer core.

    ANS: B   DIF: I   OBJ: 4-1.2

25. The strong, lower part of the mantle that lies beneath the asthenosphere is called the
    a. mesosphere.
    b. lithosphere.
    c. inner core.
    d. outer core.

    ANS: A   DIF: I   OBJ: 4-1.2

26. The soft layer of the mantle on which pieces of the lithosphere move is called the
    a. mesosphere.
    b. asthenosphere.
    c. inner core.
    d. outer core.

    ANS: B   DIF: I   OBJ: 4-1.2

27. The solid, dense center of our planet is called the
    a. mesosphere.
    b. asthenosphere.
    c. inner core.
    d. outer core.

    ANS: C   DIF: I   OBJ: 4-1.2

28. The liquid layer that lies beneath the mantle and surrounds the inner core is called the
    a. mesosphere.
    b. lithosphere.
    c. asthenosphere.
    d. outer core.

    ANS: D   DIF: I   OBJ: 4-1.2

29. The word *lithosphere* means
    a. "weak sphere."
    b. "middle sphere."
    c. "inner sphere."
    d. "rock sphere."

    ANS: D   DIF: I   OBJ: 4-1.2

30. The word *mesosphere* means
    a. "weak sphere."
    b. "middle sphere."
    c. "inner sphere."
    d. "rock sphere."

    ANS: B   DIF: I   OBJ: 4-1.2

31. The word *asthenosphere* means
    a. "weak sphere."
    b. "middle sphere."
    c. "inner sphere."
    d. "rock sphere."

    ANS: A   DIF: I   OBJ: 4-1.2

32. The ____ is made of the crust and the rigid, upper part of the mantle.
    a. mesosphere
    b. lithosphere
    c. asthenosphere
    d. outer core

    ANS: B   DIF: I   OBJ: 4-1.2

33. The ____ is made of solid rock that flows very slowly.
    a. lithosphere
    b. mesosphere
    c. asthenosphere
    d. outer core

    ANS: C   DIF: I   OBJ: 4-1.2

34. The ____ is divided into tectonic plates.
    a. mesosphere
    b. asthenosphere
    c. lithosphere
    d. outer core

    ANS: C   DIF: I   OBJ: 4-1.2

35. Which of the following is NOT a major tectonic plate?
    a. Caribbean plate
    b. Antarctic plate
    c. Eurasian plate
    d. North American plate

    ANS: A   DIF: I   OBJ: 4-1.3

36. Which of the major tectonic plates is the smallest?
    a. Australian plate
    b. Pacific plate
    c. Antarctic plate
    d. Cocos plate

    ANS: D   DIF: I   OBJ: 4-1.3

Copyright © by Holt, Rinehart and Winston. All rights reserved.

37. How many major tectonic plates are there?
    a. five
    b. seven
    c. ten
    d. fourteen

    ANS: C   DIF: I   OBJ: 4-1.3

38. The thickest part of a tectonic plate lies below
    a. the middle of the ocean.
    b. a mountain range.
    c. the center of a continent.
    d. a coastal plain.

    ANS: B   DIF: I   OBJ: 4-1.3

39. The thinnest part of a tectonic plate lies beneath
    a. the middle of the ocean.
    b. a mountain range.
    c. the center of a continent.
    d. a coastal plain.

    ANS: A   DIF: I   OBJ: 4-1.3

40. The North American plate consists
    a. entirely of oceanic crust.
    b. entirely of continental crust.
    c. of both continental and oceanic crust.
    d. None of the above

    ANS: C   DIF: II   OBJ: 4-1.3

41. Mountain ranges that occur in continental crust
    a. have very shallow roots relative to their height.
    b. have very steep roots relative to their height.
    c. have roots that are equally as deep as the mountains are high.
    d. have roots of unknown depth.

    ANS: B   DIF: I   OBJ: 4-1.3

42. The Earth's inner core has a radius of 1,228 km and the outer core is 2,200 km thick. If the Earth's radius is approximately 6,856 km, the core is ____ the Earth.
    a. a quarter of the size of
    b. a third of the size of
    c. half as big as
    d. three-quarters of the size of

    ANS: C   DIF: II   OBJ: 4-1.2

43. Which statement does NOT correctly describe seismic waves?
    a. Seismic waves travel faster through liquid than solid rock.
    b. Seismic waves cannot change speed.
    c. Scientists record seismic waves using seismographs.
    d. Scientists use seismic waves to learn about the mantle and the core.

    ANS: B   DIF: I   OBJ: 4-1.4

44. Wegener's continental drift theory suggests that at 245 million years a single landmass was surrounded by a sea called
   a. Pangaea.
   b. Gondwana.
   c. Laurasia.
   d. Panthalassa.

   ANS: D   DIF: I   OBJ: 4-2.1

45. Wegener's continental drift theory suggests that at 180 million years a single landmass gradually broke into two big pieces called
   a. Pangaea and Panthalassa.
   b. Gondwana and Laurasia.
   c. Laurasia and Eurasia.
   d. Panthalassa and Eurasia.

   ANS: B   DIF: I   OBJ: 4-2.1

46. _____ is the process by which an oceanic plate slides down the lithosphere-asthenosphere boundary.
   a. Slab pull
   b. Convection
   c. Ridge push
   d. Sea-floor spreading

   ANS: C   DIF: I   OBJ: 4-3.1

47. _____ is the process in which an oceanic plate sinks and pulls the rest of the tectonic plate with it.
   a. Slab pull
   b. Convection
   c. Ridge push
   d. Sea-floor spreading

   ANS: A   DIF: I   OBJ: 4-3.1

48. _____ is the process by which hot material from deep within the Earth rises while cooler material near the surface sinks.
   a. Slab pull
   b. Convection
   c. Ridge push
   d. Sea-floor spreading

   ANS: B   DIF: I   OBJ: 4-3.1

49. Ridge push, slab pull, and convection are all driven by heat and
   a. electromagnetism.
   b. friction.
   c. buoyancy.
   d. gravity.

   ANS: D   DIF: I   OBJ: 4-3.1

50. When two tectonic plates slide past each other horizontally, the boundary between them is called a
   a. divergent boundary.
   b. transform boundary.
   c. convergent boundary.
   d. subduction zone.

   ANS: B   DIF: I   OBJ: 4-3.2

51. When two tectonic plates move away from one another, the boundary between them is called a
   a. divergent boundary.
   b. transform boundary.
   c. convergent boundary.
   d. subduction zone.

   ANS: A   DIF: I   OBJ: 4-3.2

Copyright © by Holt, Rinehart and Winston. All rights reserved.

52. The San Andreas Fault in California is a part of the boundary between the
    a. Cocos plate and the Nazca plate.
    b. North American plate and the South American plate.
    c. Pacific plate and the Cocos plate.
    d. Pacific plate and the North American plate.

    ANS: D   DIF: I   OBJ: 4-3.2

53. Tectonic movements are generally measured in _____ per year.
    a. millimeters   c. meters
    b. centimeters   d. kilometers

    ANS: B   DIF: I   OBJ: 4-3.3

54. When vertical stress acts on a rock, _____ form.
    a. synclines   c. monoclines
    b. anticlines   d. Both (a) and (b)

    ANS: C   DIF: I   OBJ: 4-4.1

55. When horizontal stress acts on a rock, _____ form.
    a. synclines   c. monoclines
    b. anticlines   d. Both (a) and (b)

    ANS: D   DIF: I   OBJ: 4-4.1

56. A type of fault in which the hanging wall moves up relative to the footwall is called a
    a. strike-slip fault.   c. normal fault.
    b. reverse fault.   d. fault block.

    ANS: B   DIF: I   OBJ: 4-4.2

57. The type of fault that often results when rocks are pulled apart due to tension is called a
    a. strike-slip fault.   c. normal fault.
    b. reverse fault.   d. fault block.

    ANS: C   DIF: I   OBJ: 4-4.2

58. The type of fault that often results when rocks are pushed together by compression is called a
    a. strike-slip fault.   c. normal fault.
    b. reverse fault.   d. fault block.

    ANS: B   DIF: I   OBJ: 4-4.2

59. A _____ fault often results when opposing forces cause rock to break and move horizontally.
    a. strike-slip   c. normal
    b. reverse   d. block

    ANS: A   DIF: I   OBJ: 4-4.2

Copyright © by Holt, Rinehart and Winston. All rights reserved.

60. The Mariana trench is the deepest point in the oceans—11,033 m below sea level. This trench was formed at a _____ boundary, where one tectonic plate was subducted beneath the other.
    a. divergent
    b. transform
    c. convergent
    d. strike-slip

    ANS: C   DIF: II   OBJ: 4-3.2

61. Mountains with sharp, jagged peaks, such as the Tetons, in western Wyoming, that are produced when sedimentary rock layers are tilted up by faulting are called _____ mountains.
    a. folded
    b. volcanic
    c. convergent
    d. fault-block

    ANS: D   DIF: I   OBJ: 4-4.3

62. What type of mountains led early explorers to call the rim of the Pacific Ocean the Ring of Fire?
    a. folded
    b. volcanic
    c. normal
    d. fault-block

    ANS: B   DIF: I   OBJ: 4-4.3

63. What type of mountain is the only one that is formed by adding new material to the Earth's surface?
    a. folded
    b. volcanic
    c. normal
    d. fault-block

    ANS: B   DIF: I   OBJ: 4-4.4

Examine the diagram below, and answer the questions that follow.

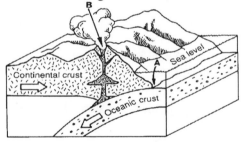

64. In the diagram above, which of the following has formed at **A**?
    a. an ocean trench
    b. a mid-ocean ridge
    c. a transform boundary
    d. none of the above

    ANS: A   DIF: II   OBJ: 4-3.2

65. In the diagram above, which of the following has formed at **B**?
    a. a folded mountain
    b. a fault-block mountain
    c. a volcanic mountain
    d. none of the above

    ANS: C   DIF: II   OBJ: 4-4.4

66. In the diagram above, which type of boundary is occurring at **A**?
    a. a divergent boundary   c. a strike-slip boundary
    b. a transform boundary   d. a convergent boundary

    ANS: D   DIF: II   OBJ: 4-3.2

67. In the diagram above, which type of tectonic plate motion is occurring at **A**?
    a. ridge push   c. convection
    b. slab pull   d. sea-floor spreading

    ANS: B   DIF: II   OBJ: 4-3.1

68. Sea-floor spreading occurs at which type of plate boundary?
    a. a divergent boundary   c. a convergent boundary
    b. a transform boundary   d. all of the above

    ANS: A   DIF: I   OBJ: 4-3.2

69. The type of collision that occurs when two lithospheric plates converge is determined primarily by the _____ of the plate.
    a. density   c. size
    b. mass   d. magnetism

    ANS: A   DIF: I   OBJ: 4-3.1

70. In which layer of the Earth are convection currents believed to occur?
    a. outer core   c. the lithosphere
    b. mesosphere   d. the asthenosphere

    ANS: D   DIF: II   OBJ: 4-3.1

71. Which of the following statements concerning magnetic reversal is NOT true?
    a. Magnetic reversals are recorded in rocks on the ocean floor.
    b. The north and south magnetic poles have changed many times throughout Earth's history.
    c. Magnetic mineral grains in rocks on the ocean floor all point in the same direction.
    d. none of the above

    ANS: C   DIF: I   OBJ: 4-2.4

72. Which of the following was NOT used as evidence to support the theory of continental drift?
    a. the existence of convection currents
    b. the similarity of fossils found on continental coasts
    c. the close fit of continental coastlines
    d. the matching of glacial grooves on different continents

    ANS: A   DIF: I   OBJ: 4-2.1

COMPLETION

1. Earth's _____ is liquid. (inner core or outer core)

    ANS: outer core   DIF: I   OBJ: 4-1.2

2. The asthenosphere is the layer of the Earth just below the _____. (mesosphere or lithosphere)

    ANS: lithosphere   DIF: I   OBJ: 4-1.2

3. The ridge in the middle of the Atlantic Ocean formed along a _____ boundary. (convergent or divergent)

    ANS: divergent   DIF: I   OBJ: 4-3.2

4. When tectonic forces push on rocks, they usually cause _____ faults. (normal or reverse)

    ANS: normal   DIF: I   OBJ: 4-4.2

5. Anticlines and synclines are the result of _____. (faults or folding)

    ANS: folding   DIF: I   OBJ: 4-4.1

6. The outermost layer of the Earth is called the _____.

    ANS: crust   DIF: I   OBJ: 4-1.1

7. The center of the Earth is called the _____.

    ANS: core   DIF: I   OBJ: 4-1.1

8. Earth's _____ crust has a composition similar to granite.

    ANS: continental   DIF: I   OBJ: 4-1.1

9. Earth's _____ crust has a composition similar to basalt.

    ANS: oceanic   DIF: I   OBJ: 4-1.1

10. The lithosphere is divided into pieces called _____.

    ANS: tectonic plates   DIF: I   OBJ: 4-1.2

11. Vibrations that travel through the Earth are called _____.

    ANS: seismic waves   DIF: I   OBJ: 4-1.4

Holt Science and Technology
Copyright © by Holt, Rinehart and Winston. All rights reserved.

12. When an earthquake occurs, _____ measure the difference in the arrival times of seismic waves and record them.

ANS: seismographs   DIF: I   OBJ: 4-1.4

13. _____ is the theory that continents can drift apart from one another and have done so in the past.

ANS: Continental drift   DIF: I   OBJ: 4-2.1

14. _____ is the process by which new oceanic lithosphere is created as older materials are pulled away.

ANS: Sea-floor spreading   DIF: I   OBJ: 4-2.2

15. _____ are underwater mountain chains that run through Earth's ocean basins.

ANS: Mid-ocean ridges   DIF: I   OBJ: 4-2.3

16. When Earth's magnetic poles change place, this is called a(n) _____ reversal.

ANS: magnetic   DIF: I   OBJ: 4-2.4

17. The region where oceanic plates sink down into the asthenosphere is called a _____.

ANS: subduction zone   DIF: I   OBJ: 4-3.1

18. _____ is the theory that the Earth's lithosphere is divided into tectonic plates that move around on top of the asthenosphere.

ANS: Plate tectonics   DIF: I   OBJ: 4-3.1

19. Scientists use a network of satellites called the _____ to measure the rate of tectonic plate movement.

ANS: Global Positioning System (GPS)

DIF: I   OBJ: 4-3.1

20. _____ is the amount of force per unit area that is put on a given material.

ANS: Stress   DIF: I   OBJ: 4-4.1

21. The stress that occurs when two tectonic plates collide is called _____.

ANS: compression   DIF: I   OBJ: 4-4.1

22. The stress that occurs when two tectonic plates pull apart from each other is called
_____.

ANS: tension   DIF: I   OBJ: 4-4.1

23. _____ occurs when rock layers bend due to stress in the Earth's crust.

ANS: Folding   DIF: I   OBJ: 4-4.1

24. The surface along which rocks break and slide past each other is called a(n)
_____.

ANS: fault   DIF: I   OBJ: 4-4.2

25. When rock changes its shape due to stress, this reaction is called _____.

ANS: deformation   DIF: I   OBJ: 4-4.1

SHORT ANSWER

For each pair of terms, explain the difference in their meanings.

1. normal fault/reverse fault

ANS:
Normal faults occur when the hanging wall moves down relative to the footwall, and reverse faults occur when the hanging wall moves up relative to the footwall.

DIF: I   OBJ: 4-4.2

2. oceanic crust/oceanic lithosphere

ANS:
Oceanic crust is crust under the oceans. Oceanic lithosphere includes oceanic crust and the rigid part of the mantle that lies below it.

DIF: I   OBJ: 4-4.2

3. folding/faulting

ANS:
Folding occurs when tectonic forces bend rock layers; faulting occurs when tectonic forces break rock.

DIF: I   OBJ: 4-4.2

4. convergent boundary/divergent boundary

ANS:
At a convergent boundary, two tectonic plates collide; at a divergent boundary, two tectonic plates pull apart.

DIF: I   OBJ: 4-3.2

5. lithosphere/asthenosphere

ANS:
Lithosphere means "rock sphere" and is the rigid outer layer of the Earth. Asthenosphere means "weak sphere" and is the soft, partially molten layer of the mantle below the lithosphere.

DIF: I   OBJ: 4-1.2

6. oceanic crust/continental crust

ANS:
Oceanic crust is a relatively thin, dense layer of crust underneath the oceans that has a composition similar to that of basalt. Continental crust is a relatively thick, lightweight layer of crust that makes up the Earth's continents and has a composition similar to that of granite.

DIF: I   OBJ: 4-1.1

7. What is the difference between continental and oceanic crust?

ANS:
Oceanic crust is thin and dense compared with continental crust. Continental crust and granite have a similar composition, and oceanic crust and basalt have a similar composition.

DIF: I   OBJ: 4-1.1

8. How is the lithosphere different from the asthenosphere?

ANS:
The lithosphere is rigid and is divided into tectonic plates. The asthenosphere is a layer of soft mantle material that flows very slowly.

DIF: I   OBJ: 4-1.2

9. How do scientists know about the structure of the Earth's interior? Explain.

ANS:
Scientists measure the different speeds at which seismic waves travel through different parts of the Earth. This indicates the density and thickness of each layer the waves pass through.

DIF: I   OBJ: 4-1.4

Copyright © by Holt, Rinehart and Winston. All rights reserved.

10. Explain the difference between the crust and the lithosphere.

ANS:
The crust and the lithosphere are the outermost layers of the Earth, but the lithosphere includes the crust and the rigid, uppermost part of the mantle.

DIF: II   OBJ: 4-1.2

11. List three puzzling occurrences that the theory of continental drift helped to explain, and describe how it explained them.

ANS:
Occurrences include the puzzlelike fit of the continents, the match of glacial grooves, the occurrence of fossils of the same species on different continents, the distribution of rock types and ancient climatic zones. Continental drift explained that these coincidences exist because at one time all the continents were joined together in one large landmass.

DIF: I   OBJ: 4-2.1

12. Explain why Wegener's theory of continental drift was not accepted at first.

ANS:
Wegener's theory of continental drift described the movement of continents but did not explain what forces of nature moved them.

DIF: I   OBJ: 4-2.1

13. Explain how the processes of sea-floor spreading and magnetic reversal produce bands of oceanic crust that have different magnetic polarities.

ANS:
During sea-floor spreading, new oceanic crust forms on either side of the mid-ocean ridge. The changing polarity of the Earth's magnetic poles causes the new oceanic crust to have alternating bands of normal and reverse polarity.

DIF: II   OBJ: 4-2.4

14. List and describe three possible driving forces of tectonic plate motion.

ANS:
Ridge push occurs when an oceanic plate slides down the tilted slope of the lithosphere/asthenosphere boundary. Slab pull occurs when the sinking edge of an oceanic plate pulls the rest of the plate down with it into the subduction zone. Convection occurs when hot mantle material in the asthenosphere convects, dragging the tectonic plate sideways.

DIF: I   OBJ: 4-3.1

15. How do the three types of convergent boundaries differ from one another?

ANS:
Convergent boundaries can occur between two oceanic plates, two continental plates, or between an oceanic and a continental plate.

DIF: I   OBJ: 4-3.2

16. Explain how scientists measure the rate at which tectonic plates move.

ANS:
They measure tectonic plate movement by using a network of satellites to track the movement of GPS ground stations over long periods of time.

DIF: I   OBJ: 4-3.3

17. When convection takes place in the mantle, why does cooler material sink, while warmer material rises?

ANS:
Cooler material sinks because it is denser than warmer material.

DIF: II   OBJ: 4-3.1

18. What is the difference between an anticline and a syncline?

ANS:
An anticline is shaped like an upside-down bowl, while a syncline resembles a bowl that is right-side up.

DIF: I   OBJ: 4-4.1

19. What is the difference between a normal fault and a reverse fault?

ANS:
In a normal fault, the hanging wall moves down relative to the footwall. In a reverse fault, the hanging wall moves up relative to the footwall.

DIF: I   OBJ: 4-4.2

20. Name and describe the type of tectonic stress that forms folded mountains.

ANS:
Folded mountains form when compression acts on rock strata (such as when two continental plates collide) so that the layers of rock are pushed up into huge folds.

DIF: I   OBJ: 4-4.4

Copyright © by Holt, Rinehart and Winston. All rights reserved.

21. Name and describe the type of tectonic stress that forms fault-block mountains.

ANS:
Fault-block mountains form when tension pulls rock apart, causing a large number of normal faults to form. When some of these fault blocks drop down relative to others, fault-block mountains form.

DIF: I   OBJ: 4-4.4

22. If a fault occurs in an area where rock layers have been folded, which type of fault is it likely to be? Why?

ANS:
A reverse fault is likely to form because both reverse faults and folding occur in areas where compression takes place.

DIF: II   OBJ: 4-4.2

23. If the Earth's crust is growing at mid-ocean ridges, why doesn't the Earth itself grow larger?

ANS:
The Earth doesn't grow larger because the crust is part of the rock cycle.

DIF: I   OBJ: 4-3.1

24. What was Pangaea?

ANS:
Pangaea was the large landmass that later broke up to form two supercontinents and then fragmented further to form the six continents of today.

DIF: I   OBJ: 4-2.1

25. Why are there several categories of convergent plate boundaries?

ANS:
Plates that are pushed together behave differently, depending on their composition and density.

DIF: I   OBJ: 4-3.2

26. Where would you expect to see the following features?
    a.   tall, wrinkled mountains in the middle of a continent
    b.   a long parallel ridge on the ocean floor surrounded by parallel zones of magnetic reversal

ANS:
    a.   I would expect to see this at a convergent continental/continental boundary.
    b.   I would expect to see this at a divergent boundary.

DIF: I   OBJ: 4-3.2

27. Explain the process of subduction.

ANS:
A denser oceanic plate is forced beneath a less-dense oceanic or continental plate at a convergent boundary. Gravity pulls the oceanic plate in to the asthenosphere, where it begins to melt.

DIF: I   OBJ: 4-3.2

28. What three features form when rock layers bend?

ANS:
When rock layers bend, anticlines, synclines, and monoclines form.

DIF: I   OBJ: 4-4.1

29. Why are the Appalachian Mountains now located in the middle of the North American Plate?

ANS:
The Appalachians formed when North America and Africa collided. In time, the plates separated and so much new crust was created that the mountains were no longer located at the plate boundary.

DIF: I   OBJ: 4-4.4

30. What is a tectonic plate?

ANS:
A tectonic plate is a large piece of the lithosphere that moves around on top of the asthenosphere.

DIF: I   OBJ: 4-1.3

31. What was the major problem with Wegener's theory of continental drift?

ANS:
Wegener's theory did not explain the driving force responsible for continental drift.

DIF: I   OBJ: 4-2.1

32. Why is there stress on the Earth's crust?

ANS:
Stress occurs in the Earth's crust because the crust is part of all tectonic plates, and tectonic plates are constantly colliding, pulling apart, and sliding past each other.

DIF: I   OBJ: 4-4.1

Copyright © by Holt, Rinehart and Winston. All rights reserved.

33. Use the following terms to create a concept map: *sea-floor spreading, convergent boundary, divergent boundary, subduction zone, transform boundary, tectonic plates.*

ANS:

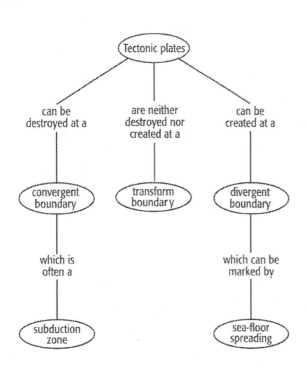

DIF: I   OBJ: 4-3.2

34. Why is it necessary to think about the different layers of the Earth in terms of both their composition and their physical properties?

ANS:
Some layers of the Earth (such as the inner and outer cores) have the same composition but different physical properties.

DIF: II   OBJ: 4-1.2

35. Folded mountains usually form at the edge of a tectonic plate. How can you explain old folded mountain ranges located in the middle of a tectonic plate?

ANS:
At the time they formed, the folded mountains must have been on the edge of a tectonic plate. New material was later added to the tectonic plate, causing the folded mountains to be closer to the center of the plate.

DIF: II   OBJ: 4-4.4

36. New tectonic plate material continually forms at divergent boundaries. Tectonic plate material is also continually destroyed in subduction zones at convergent boundaries. Do you think the total amount of lithosphere formed on Earth is about equal to the amount destroyed? Why?

ANS:
Answers will vary. Sample answer: The amount of crust formed is roughly equal to the amount of crust destroyed globally. If this were not true, the Earth would be either expanding or shrinking.

DIF: II   OBJ: 4-3.2

37. Assume that a very small oceanic plate is between a mid-ocean ridge to the west and a subduction zone to the east. At the ridge, the oceanic plate is growing at a rate of 5 km every million years. At the subduction zone, the oceanic plate is being destroyed at a rate of 10 km every million years. If the oceanic plate is 100 km across, in how many million years will the oceanic plate disappear?

ANS:
In 1 million years, the tectonic plate grows 5 km on one side but shrinks by 10 km on the other side. Every 1 million years, the tectonic plate shrinks by 5 km. In 20 million years, the tectonic plate will disappear entirely. The rate of tectonic plate destruction is 5 km/y – 10 km/y = –5 km/y. The tectonic plate will completely disappear in 100 km ÷ 5 km/y = 20 million years.

DIF: II   OBJ: 4-2.3

Imagine that you could travel to the center of the Earth. Use the table below to answer the questions that follow.

Composition	Structure
Crust (50 km)	Lithosphere (150 km)
Mantle (2,900 km)	Asthenosphere (250 km)
	Mesosphere (2,550 km)
Core	Outer core (2,200 km)
(3,428 km)	Inner core (1,228 km)

38. How far beneath Earth's surface would you have to go to find the liquid material in the Earth's core?

ANS:
150 km + 250 km + 2,550 km = 2,950 km

DIF: II   OBJ: 4-1.2

39. At what range of depth would you find mantle material but still be within the lithosphere?

ANS:
You would find mantle material in the lithosphere between 50 and 150 km.

DIF: II   OBJ: 4-1.2

40. Explain how folded mountain ranges form.

ANS:
Folded mountain ranges form when two tectonic plates with continental crust collide. As the continents crash into each other, the crust buckles and thickens at the points of collisions. This causes the continental crust to be pushed upward, which forms mountains over a long period of time.

DIF: I   OBJ: 4-4.4

41. Describe the role of the asthenosphere in the movement of tectonic plates.

ANS:
The asthenosphere is a layer made of solid rock that flows very slowly. The tectonic plates move very slowly on top of the asthenosphere.

DIF: I   OBJ: 4-1.3

42. Use the table below to answer the following question.

Seismic Waves in Material Densities

Material	Speed of earthquake wave (km/s)
Sediment	2.0
Granite	6.0
Basalt	7.0
Peridotite	8.0

An earthquake wave reached a seismic station 2,080 km from the focus in 347 seconds. Through which of the above materials did the wave travel? Show your work.

ANS:
2,080 km/347 s = 6 km/s, which is the speed of a wave traveling through granite.

DIF: II   OBJ: 4-1.4

43. Give evidence that sea-floor spreading exists.

    ANS:
    Sea-floor spreading is the process by which new oceanic crust is created as older materials are pushed or pulled away. Magma cools and forms new rocks in mid-ocean ridges. Because rocks age as they move away from the ridges, rocks near the edges of the continents are much older than rocks near mid-ocean ridges. This is evidence of sea-floor spreading.

    DIF: II   OBJ: 4-2.2

    Examine the diagrams of convergent boundaries and answer the questions that follow.

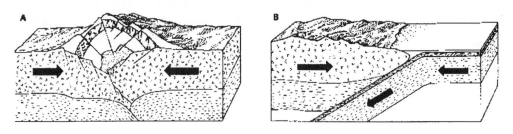

44. Which type of tectonic plates are colliding in A? Explain.

    ANS:
    A shows two continental plates colliding. The continental crust is pushed upwards.

    DIF: II   OBJ: 4-1.3

45. Which two types of tectonic plates are colliding in B? Explain.

    ANS:
    A continental plate is colliding with an oceanic plate in B. The denser oceanic plate is pushed beneath the continental plate.

    DIF: II   OBJ: 4-1.3

46. Which diagram represents the type of boundary that creates some of the world's tallest mountains? Explain.

    ANS:
    Diagram A represents a continental/continental boundary, which creates mountains by pushing the continental crust upward.

    DIF: II   OBJ: 4-3.2

Copyright © by Holt, Rinehart and Winston. All rights reserved.

47. Use the following terms to complete the concept map below: *South America, Panthalassa, Gondwana, Laurasia, North America, Pangaea.*

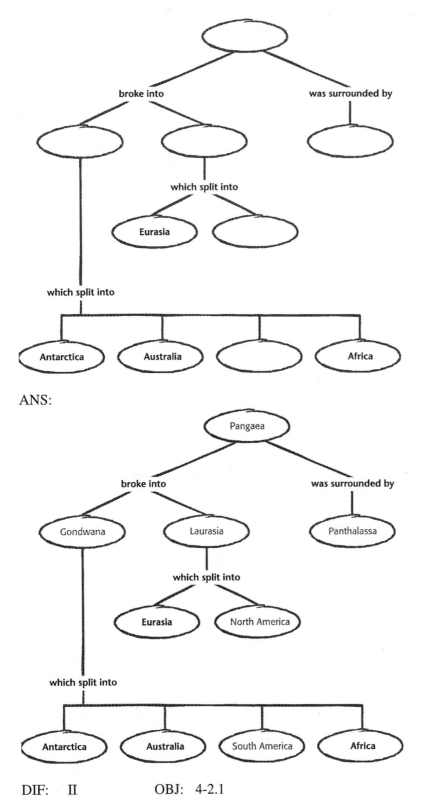

ANS:

DIF: II   OBJ: 4-2.1

Copyright © by Holt, Rinehart and Winston. All rights reserved.

MULTIPLE CHOICE

1. When rock is ____, energy builds up in it. Seismic waves occur as this energy is
   a. elastically deformed, released.
   b. plastically deformed, released.
   c. elastically deformed, increased.
   d. plastically deformed, increased.

   ANS: A   DIF: I   OBJ: 5-1.1

2. The strongest earthquakes usually occur
   a. near divergent boundaries.
   b. near convergent boundaries.
   c. near transform boundaries.
   d. along normal faults.

   ANS: B   DIF: I   OBJ: 5-1.2

3. The last seismic waves to arrive are
   a. P waves.
   b. S waves.
   c. surface waves.
   d. body waves.

   ANS: C   DIF: I   OBJ: 5-1.3

4. If an earthquake begins while you are in a building, the safest thing to do first is
   a. get under the strongest table, chair, or other piece of furniture.
   b. run out into the street.
   c. crouch near a wall.
   d. call home.

   ANS: A   DIF: I   OBJ: 5-3.4

5. Studying earthquake waves currently allows seismologists to do all of the following EXCEPT
   a. determine when an earthquake started.
   b. learn about the Earth's interior.
   c. decrease an earthquake's strength.
   d. determine where an earthquake started.

   ANS: C   DIF: I   OBJ: 5-2.3

6. If a planet has a liquid core, then S waves
   a. speed up as they travel through the core.
   b. maintain their speed as they travel through the core.
   c. change direction as they travel through the core.
   d. cannot pass through the core.

   ANS: D   DIF: I   OBJ: 5-1.3

7. Strike-slip faults are prominent along ____ plate boundaries
   a. convergent
   b. transform
   c. transcontinental
   d. divergent

   ANS: B   DIF: I   OBJ: 5-1.2

Copyright © by Holt, Rinehart and Winston. All rights reserved.

8. What causes the ground to move during an earthquake?
   a. elastic rebound
   b. deformation
   c. stress
   d. tectonic force

   ANS: A   DIF: I   OBJ: 5-1.1

9. Primary seismic waves
   a. are slower than secondary waves.
   b. are the result of shearing forces in rock.
   c. can travel through solids, liquids, and gases.
   d. cause the Earth's surface to roll up and down.

   ANS: C   DIF: I   OBJ: 5-1.3

10. Most earthquakes occur along or near the edges of the Earth's
    a. oceans.
    b. tectonic plates.
    c. rivers.
    d. continents.

    ANS: B   DIF: I   OBJ: 5-1.1

11. _____ motion occurs where two tectonic plates slip past each other.
    a. Convergent
    b. Divergent
    c. Transform
    d. all of the above

    ANS: C   DIF: I   OBJ: 5-1.2

12. _____ motion occurs where two tectonic plates push together.
    a. Convergent
    b. Divergent
    c. Transform
    d. all of the above

    ANS: A   DIF: I   OBJ: 5-1.2

13. _____ motion occurs where two tectonic plates pull away from each other.
    a. Convergent
    b. Divergent
    c. Transform
    d. all of the above

    ANS: B   DIF: I   OBJ: 5-1.2

14. Shallow, moderate earthquakes are produced along _____ faults.
    a. normal
    b. converse
    c. reverse
    d. strike-slip

    ANS: D   DIF: I   OBJ: 5-1.2

15. Strong, deep earthquakes are produced along _____ faults.
    a. strike-slip
    b. converse
    c. reverse
    d. normal

    ANS: C   DIF: I   OBJ: 5-1.2

Copyright © by Holt, Rinehart and Winston. All rights reserved.

16. Weak, shallow earthquakes are produced along _____ faults.
    a. normal   c. reverse
    b. strike-slip   d. converse

    ANS: A   DIF: I   OBJ: 5-1.2

17. _____ motion causes a reverse fault.
    a. Convergent   c. Transform
    b. Divergent   d. all of the above

    ANS: A   DIF: I   OBJ: 5-1.2

18. _____ motion causes a normal fault.
    a. Convergent   c. Transform
    b. Divergent   d. all of the above

    ANS: B   DIF: I   OBJ: 5-1.2

19. The footwall moves down relative to the hanging wall in _____ faults.
    a. normal   c. reverse
    b. converse   d. strike-slip

    ANS: C   DIF: I   OBJ: 5-1.2

20. The footwall moves up relative to the hanging wall in _____ faults.
    a. normal   c. reverse
    b. converse   d. strike-slip

    ANS: A   DIF: I   OBJ: 5-1.2

21. P waves and S waves are two types of _____ waves.
    a. interior   c. surface
    b. exterior   d. body

    ANS: D   DIF: I   OBJ: 5-1.3

22. Which type of seismic wave travels the fastest?
    a. an L wave   c. an S wave
    b. a P wave   d. a surface wave

    ANS: B   DIF: I   OBJ: 5-1.3

23. Which type of seismic wave can travel through solids, liquids, and gases?
    a. an L wave   c. an S wave
    b. a P wave   d. a surface wave

    ANS: B   DIF: I   OBJ: 5-1.3

Copyright © by Holt, Rinehart and Winston. All rights reserved.

24. Which type of seismic wave cannot travel through liquids?
    a. an L wave
    c. an S wave
    b. a P wave
    d. a surface wave

    ANS: C   DIF: I   OBJ: 5-1.3

25. Which type of seismic wave is also called a secondary wave?
    a. an L wave
    c. an S wave
    b. a P wave
    d. a surface wave

    ANS: C   DIF: I   OBJ: 5-1.3

26. Which type of seismic wave is also called a primary wave?
    a. an L wave
    c. an S wave
    b. a P wave
    d. a surface wave

    ANS: B   DIF: I   OBJ: 5-1.3

27. Which type of seismic wave travels the slowest?
    a. an L wave
    c. an S wave
    b. a P wave
    d. a surface wave

    ANS: D   DIF: I   OBJ: 5-1.3

28. Which type of seismic wave causes a shearing effect?
    a. an L wave
    c. an S wave
    b. a P wave
    d. a surface wave

    ANS: C   DIF: I   OBJ: 5-1.3

29. Which type of seismic wave is the most destructive?
    a. an L wave
    c. an S wave
    b. a P wave
    d. a surface wave

    ANS: D   DIF: I   OBJ: 5-1.3

Examine the illustration below, and answer the questions that follow.

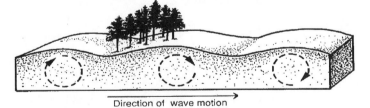

Direction of wave motion

30. Which type of seismic wave is illustrated above?
    a. a surface wave
    c. an S wave
    b. a P wave
    d. a body wave

    ANS: A   DIF: II   OBJ: 5-1.3

Holt Science and Technology
Copyright © by Holt, Rinehart and Winston. All rights reserved.

31. Seismologists find an earthquake's start time by comparing seismograms and noting the difference in arrival times of
    a. body waves and surface waves.
    b. P waves and S waves.
    c. S waves and surface waves.
    d. P waves and surface waves.

    ANS: B   DIF: I   OBJ: 5-2.1

Examine the illustration below and answer the questions that follow.

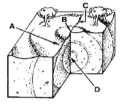

32. Which point in the illustration represents the epicenter of the earthquake?
    a. point **A**
    b. point **B**
    c. point **C**
    d. point **D**

    ANS: B   DIF: II   OBJ: 5-2.1

33. Which point in the illustration represents the focus of the earthquake?
    a. point **A**
    b. point **B**
    c. point **C**
    d. point **D**

    ANS: D   DIF: II   OBJ: 5-2.1

34. What is the minimum number of seismograph stations that is necessary to locate the epicenter of an earthquake?
    a. one
    b. two
    c. three
    d. four

    ANS: C   DIF: I   OBJ: 5-2.2

35. If a seismograph detects S waves shortly after it detects P waves, then the earthquake was
    a. nearby.
    b. far away.
    c. very weak.
    d. very strong.

    ANS: A   DIF: I   OBJ: 5-2.2

36. The Richter scale is used to measure the
    a. length of time an earthquake lasts.
    b. the epicenter of an earthquake.
    c. strength of an earthquake.
    d. depth of an earthquake's focus.

    ANS: C   DIF: I   OBJ: 5-2.3

37. Damage occurs at the epicenter when the magnitude of an earthquake reaches ____ on the Richter scale.
    a. 4.0
    b. 5.0
    c. 6.0
    d. 7.0

    ANS: B   DIF: I   OBJ: 5-2.3

Holt Science and Technology
Copyright © by Holt, Rinehart and Winston. All rights reserved.
100

38. How much more energy is released by an earthquake with a magnitude of 3.0 than by an earthquake with a magnitude of 2.0?
   a. 1,000 times more energy
   b. 317 times more energy
   c. 100 times more energy
   d. 31.7 times more energy

   ANS: D   DIF: I   OBJ: 5-2.3

39. The magnitude of an earthquake is a direct measure of
   a. how much energy it releases.
   b. how much damage it causes.
   c. how long it lasts.
   d. how many aftershocks it causes.

   ANS: A   DIF: I   OBJ: 5-2.3

40. About how much more energy is released by an earthquake with a magnitude of 4.0 than by an earthquake with a magnitude of 2.0?
   a. twice as much energy
   b. 31.7 times more energy
   c. 63.4 times more energy
   d. 1,005 times more energy

   ANS: D   DIF: II   OBJ: 5-2.3

41. Which area of the United States has the highest earthquake hazard level?
   a. the East Coast
   b. the Gulf Coast
   c. the West Coast
   d. the Midwest

   ANS: C   DIF: I   OBJ: 5-3.1

42. Which of the following helped scientists predict the 1989 Loma Prieta earthquake?
   a. peculiarities in animal behavior
   b. depths of past earthquake foci
   c. intensities of past earthquakes
   d. seismic gaps in the San Andreas Fault

   ANS: D   DIF: I   OBJ: 5-3.2

43. A(n) _____ is a weight placed in the roof of a building that moves to counteract the building's movement during an earthquake.
   a. active tendon system
   b. mass damper
   c. base isolator
   d. cross-brace

   ANS: B   DIF: I   OBJ: 5-3.3

44. A(n) _____ uses sensors and a computer to activate devices that shift a large weight at the base of a building to counteract the building's movement during an earthquake.
   a. active tendon system
   b. cross-brace
   c. base isolator
   d. mass damper

   ANS: A   DIF: I   OBJ: 5-3.3

45. _____ are placed between floors to counteract pressure that pushes and pulls at the side of a building during an earthquake.
   a. Flexible pipes
   b. Mass dampers
   c. Base isolators
   d. Cross-braces

   ANS: D   DIF: I   OBJ: 5-3.3

Holt Science and Technology
Copyright © by Holt, Rinehart and Winston. All rights reserved.

46. _____ help prevent water and gas lines from breaking during an earthquake.
    a. Flexible pipes
    b. Mass dampers
    c. Base isolators
    d. Cross-braces

    ANS: A   DIF: I   OBJ: 5-3.3

47. _____ absorb seismic waves during an earthquake, preventing them from traveling through the building.
    a. Mass dampers
    b. Flexible pipes
    c. Base isolators
    d. Cross-braces

    ANS: C   DIF: I   OBJ: 5-3.3

48. If you are outside when an earthquake begins, you should
    a. run into the nearest building.
    b. lie face down away from buildings.
    c. crouch near a wall.
    d. run away as fast as possible.

    ANS: B   DIF: I   OBJ: 5-3.4

49. If you are in a car on an open road when an earthquake begins, you should
    a. stop the car and remain inside.
    b. stop the car and get out as fast as possible.
    c. continue driving until you believe the danger is over.
    d. speed up to get away from the earthquake.

    ANS: A   DIF: I   OBJ: 5-3.4

50. The _____ marks the boundary between the Earth's crust and mantle, where the speed of seismic waves increases sharply.
    a. Moho
    b. seismic gap
    c. shadow zone
    d. subduction zone

    ANS: A   DIF: I   OBJ: 5-4.1

51. The _____ is an area on the Earth's surface where no direct seismic waves from a particular earthquake can be detected.
    a. Moho
    b. seismic gap
    c. shadow zone
    d. subduction zone

    ANS: C   DIF: I   OBJ: 5-4.1

52. The discovery of the _____ suggested that the Earth has a liquid core.
    a. Moho
    b. seismic gap
    c. shadow zone
    d. subduction zone

    ANS: C   DIF: I   OBJ: 5-4.1

53. The first and perhaps the most successful seismic test on another cosmic body was on
    a. Jupiter.
    b. Mars.
    c. the sun.
    d. the moon.

    ANS: D   DIF: I   OBJ: 5-4.2

54. After astronauts crashed a landing vehicle on the moon in 1969, a seismograph registered a seismic disturbance that lasted for
    a. 20 to 30 seconds.
    b. 5 minutes.
    c. half an hour.
    d. over an hour and a half.

    ANS: D DIF: I OBJ: 5-4.2

55. Which of the following statements best describes current scientific thought concerning the formation of Earth's moon?
    a. Earth's moon was an asteroid captured by Earth's gravitational field.
    b. Earth's moon formed when a Mars-sized object collided with Earth.
    c. Earth's moon was a comet captured by Earth's gravitational field.
    d. Earth's moon was once a part of Mars that was captured by Earth's gravitational field.

    ANS: B DIF: II OBJ: 5-4.2

56. When scientists tried to use a seismograph on Mars, _____ interfered with the machine.
    a. dust
    b. water
    c. wind
    d. solar flares

    ANS: C DIF: I OBJ: 5-4.2

57. Scientists study seismic waves on the sun using a satellite called
    a. *SOHO.*
    b. *SSWWS.*
    c. *Viking 1.*
    d. GPS.

    ANS: A DIF: I OBJ: 5-4.2

58. Scientists studying the sun have discovered that _____ produce seismic waves causing "sunquakes."
    a. tectonic plates
    b. solar flares
    c. faults
    d. convergent motions

    ANS: B DIF: I OBJ: 5-4.2

In January of 1994, an earthquake in California measuring 6.9 on the Richter scale caused the Santa Susana Mountains, just north of Los Angeles, to increase in height by 70 centimeters.

59. Which type of motion would most probably have produced this earthquake?
    a. transform motion
    b. convergent motion
    c. divergent motion
    d. all of the above

    ANS: A DIF: II OBJ: 5-1.2

60. Along which type of fault would this earthquake have most probably taken place?
    a. a normal fault
    b. a strike-slip fault
    c. a reverse fault
    d. all of the above

    ANS: B DIF: II OBJ: 5-1.2

Copyright © by Holt, Rinehart and Winston. All rights reserved.

61. What type of damage would you expect to have occurred with an earthquake of this intensity?
    a. no damage; it would not even have been felt
    b. no damage; only felt slight shaking at the epicenter
    c. only damage at the epicenter
    d. widespread damage

    ANS: D   DIF: II   OBJ: 5-3.2

In June of 1994 in Northern Bolivia, an extremely deep earthquake—at a depth of 637 km—was felt all the way from South America to Canada.

62. Which type of motion would most probably have produced this earthquake?
    a. transform motion   c. divergent motion
    b. convergent motion   d. all of the above

    ANS: B   DIF: II   OBJ: 5-1.2

63. Along which type of plate boundary would this earthquake have likely taken place?
    a. a transform boundary   c. a convergent boundary
    b. a divergent boundary   d. all of the above

    ANS: C   DIF: II   OBJ: 5-1.2

Examine the table below, and answer the questions that follow.

Worldwide Earthquake Frequency
(Based on Observations Since 1900)

Descriptor	Magnitude	Average occurring annually
Great	8.0 and higher	1
Major	7.0–7.9	18
Strong	6.0–6.9	120
Moderate	5.0–5.9	800
Light	4.0–4.9	about 6,200
Minor	3.0–3.9	about 49,000
Very minor	2.0–2.9	about 365,000

64. Look at the trends of earthquakes in the table above. Generally, based on observations of worldwide earthquakes since 1900, with each step down in earthquake magnitude, the number of earthquakes per year is about
    a. 50 to 100 times less.   c. 50 to 100 times greater.
    b. 5 to10 times less.   d. 5 to 10 times greater.

    ANS: D   DIF: II   OBJ: 5-3.2

Holt Science and Technology
Copyright © by Holt, Rinehart and Winston. All rights reserved.

65. In January of 2001, Kutch, in the Indian state of Gujarat, was the epicenter of a 30-second earthquake that registered 6.9 on the Richter scale. Using the table above, this earthquake would be described as
    a. minor.
    b. light.
    c. strong.
    d. great.

    ANS: C   DIF: II   OBJ: 5-3.2

66. According to the table, earthquakes that register between 6.0 and 6.9 on the Richter scale occur approximately _____ times a year worldwide.
    a. 18
    b. 120
    c. 800
    d. 6,200

    ANS: B   DIF: II   OBJ: 5-3.2

67. In June 1994, an earthquake in Northern Bolivia that measured 8.2 on the Richter scale was felt all the way from South America to Canada. According to the table, what type of damage near the epicenter would have occurred with an earthquake of this intensity?
    a. minor damage
    b. light damage
    c. strong damage
    d. great damage

    ANS: D   DIF: II   OBJ: 5-3.2

68. How often does an earthquake measuring 8.2 on the Richter scale, such as the one in Bolivia in 1994, occur worldwide?
    a. once every ten years
    b. once a year
    c. twice a year
    d. eight times a year

    ANS: B   DIF: II   OBJ: 5-1.2

COMPLETION

1. Energy is released as _____ occurs. (deformation or elastic rebound)

    ANS: elastic rebound   DIF: I   OBJ: 5-1.1

2. _____ cannot travel through parts of the Earth that are completely liquid. (S waves or P waves)

    ANS: S waves   DIF: I   OBJ: 5-1.3

3. Seismologists use the S-P-time method to find an earthquake's _____. (shadow zone or epicenter)

    ANS: epicenter   DIF: I   OBJ: 5-2.1

4. The _____ is a place that marks a sharp increase in seismic wave speed. (seismic gap or Moho)

    ANS: Moho   DIF: I   OBJ: 5-4.1

Holt Science and Technology
Copyright © by Holt, Rinehart and Winston. All rights reserved.

5. The instrument used to record earthquake waves is a _____. (seismograph or seismogram)

   ANS: seismograph   DIF: I   OBJ: 5-2.1

6. The _____ of an earthquake may be far below the Earth's surface. (epicenter or focus)

   ANS: focus   DIF: I   OBJ: 5-2.1

7. Sections along an active fault may have _____, where there is little earthquake activity. (seismic gaps or elastic rebound)

   ANS: seismic gaps   DIF: I   OBJ: 5-3.2

8. Secondary waves are blocked by the Earth's outer core, creating a _____ at locations on the opposite side of the Earth from the earthquake's focus. (fault or shadow zone)

   ANS: shadow zone   DIF: I   OBJ: 5-4.1

9. The bending of rock causes _____. (Moho or deformation)

   ANS: deformation   DIF: I   OBJ: 5-1.1

10. The study of earthquakes is called _____.

    ANS: seismology   DIF: I   OBJ: 5-1.1

11. The scientists who study earthquakes are called _____.

    ANS: seismologists   DIF: I   OBJ: 5-1.1

12. _____ are the giant masses of solid rock that make up the outermost part of the Earth.

    ANS: Tectonic plates   DIF: I   OBJ: 5-1.1

13. A(n) _____ is a break in the Earth's crust along which blocks of the crust slide relative to one another.

    ANS: fault   DIF: I   OBJ: 5-1.1

14. _____ deformation leads to earthquakes.

    ANS: Elastic   DIF: I   OBJ: 5-1.1

15. _____ is the sudden return of elastically deformed rock to its original shape.

    ANS: Elastic rebound   DIF: I   OBJ: 5-1.1

16. _____ waves are waves of energy that travel through the Earth.

    ANS: Seismic   DIF: I   OBJ: 5-1.3

17. _____ waves are seismic waves that travel through the Earth's interior.

    ANS: Body   DIF: I   OBJ: 5-1.3

18. _____ waves are seismic waves that travel along the Earth's surface.

    ANS: Surface   DIF: I   OBJ: 5-1.3

19. A(n) _____ is a tracing of earthquake motion created by a seismograph.

    ANS: seismogram   DIF: I   OBJ: 5-2.1

20. A(n) _____ is the point on the Earth's surface directly above an earthquake's starting point.

    ANS: epicenter   DIF: I   OBJ: 5-2.1

21. A(n) _____ is the point inside the Earth where an earthquake begins.

    ANS: focus   DIF: I   OBJ: 5-2.1

22. _____ measures how prone an area is to experiencing earthquakes in the future.

    ANS: Earthquake hazard   DIF: I   OBJ: 5-3.1

23. The _____ states that sections of active faults that have had relatively few earthquakes are likely to be the sites of strong earthquakes in the future.

    ANS: gap hypothesis   DIF: I   OBJ: 5-3.2

24. _____ are weaker earthquakes that follow stronger earthquakes.

    ANS: Aftershocks   DIF: I   OBJ: 5-3.2

25. _____ are powerful magnetic disturbances in the sun.

    ANS: Solar flares   DIF: I   OBJ: 5-4.2

SHORT ANSWER

1. Where do earthquakes occur?

   ANS:
   Answers may vary. Most earthquakes take place near the edges of tectonic plates.

   DIF: I   OBJ: 5-1.1

2. What directly causes earthquakes?

   ANS:
   Answers may vary. Most earthquakes occur when rock releases energy during elastic rebound.

   DIF: I   OBJ: 5-1.1

3. Arrange the types of earthquakes caused by the three plate-motion types from weakest to strongest.

   ANS:
   Earthquakes caused by divergent motion are weakest; earthquakes caused by transform motion are moderate; and earthquakes caused by convergent motion are strongest.

   DIF: I   OBJ: 5-1.2

4. Why are surface waves more destructive to buildings than P waves or S waves?

   ANS:
   As surface waves travel across the surface, they move the ground up and down in circles. Surface waves are more destructive because they cause the ground to move more than P waves and S waves do.

   DIF: II   OBJ: 5-1.3

5. What is the difference between a seismogram and a seismograph?

   ANS:
   A seismograph is an instrument that records seismic waves. A seismogram, which is created by a seismograph, is a tracing of the ground's motion during an earthquake.

   DIF: I   OBJ: 5-2.1

Copyright © by Holt, Rinehart and Winston. All rights reserved.

6. How many seismograph stations are needed to use the S-P-time method? Why?

ANS:
Three; two seismograph stations can narrow the location of the earthquake to two possible locations. Adding a third enables scientists to determine which of the two locations is correct.

DIF: I   OBJ: 5-2.2

7. If the amount of energy released by an earthquake with a magnitude of 7.0 on the Richter scale is x, what is the amount of energy released by an earthquake with a magnitude of 6.0 in terms of x?

ANS:
$x / 31.7$

DIF: II   OBJ: 5-2.3

8. How is an area's earthquake hazard determined?

ANS:
It is determined by past and present seismic activity.

DIF: I   OBJ: 5-3.1

9. Which earthquake forecast predicts a more precise location—a forecast based on the relationship between strength and frequency or a forecast based on the gap hypothesis?

ANS:
A forecast based on the gap hypothesis would predict a more precise location.

DIF: I   OBJ: 5-3.2

10. Describe two ways that buildings are reinforced against earthquakes.

ANS:
Describing any two features from this section (mass damper, cross braces, active tendon system, flexible pipes, and base isolators) is acceptable. Students may also describe features they learn about through additional research.

DIF: I   OBJ: 5-3.3

11. Name four items that you should store in case of an earthquake.

ANS:
Answers will vary. Students may list some items not mentioned in the textbook. Items mentioned in this section include nonperishable food, water, a fire extinguisher, a flashlight with batteries, and a first-aid kit.

DIF: I   OBJ: 5-3.4

Copyright © by Holt, Rinehart and Winston. All rights reserved.

12. What observation of seismic-wave travel led to the discovery of the Moho?

ANS:
Seismologists found a sharp increase in the speed of seismic waves and the bending of seismic waves at the Moho boundary.

DIF: I   OBJ: 5-4.1

13. Briefly describe one discovery seismologists have made about each of the following cosmic bodies: the moon, Mars, and the sun.

ANS:
Answers will vary. Possible answers include:
- Seismic waves last a lot longer on the moon than they do on the Earth.
- Mars is not very seismically active.
- "Sunquakes" are generally a lot stronger than earthquakes.

DIF: I   OBJ: 5-4.2

14. Why don't S waves enter the Earth's outer core?

ANS:
S waves do not enter the outer core because they cannot travel through parts of the Earth that are liquid.

DIF: I   OBJ: 5-1.3

15. What is a fault?

ANS:
A fault is a break in the Earth's crust along which blocks of the crust slide relative to one another.

DIF: I   OBJ: 5-1.1

16. Name two ways in which rock along a fault deforms in response to stress.

ANS:
Rock along a fault can deform in a plastic manner, like a piece of molded clay, or in an elastic manner, like a rubber band.

DIF: I   OBJ: 5-1.1

17. How is an earthquake's epicenter related to its focus?

ANS:
The epicenter is the point on the Earth's surface directly above the focus, which is where the earthquake originates.

DIF: I   OBJ: 5-2.1

18. As seismic waves travel farther, what happens to the difference in arrival times of P and S waves?

ANS:
The difference in arrival times of P and S waves increases as seismic waves travel farther.

DIF: I   OBJ: 5-2.2

19. What is the gap hypothesis?

ANS:
The gap hypothesis states that sections of active faults that have had relatively few earthquakes are likely to be the sites of strong earthquakes in the future.

DIF: I   OBJ: 5-3.2

20. Why should you lie under a table or desk during an earthquake?

ANS:
The table or desk might prevent falling objects from hitting you and causing injury.

DIF: I   OBJ: 5-3.4

21. What are aftershocks?

ANS:
Aftershocks are weaker earthquakes that follow stronger earthquakes.

DIF: I   OBJ: 5-3.2

22. Name three features in the Earth's interior that were discovered by studying seismic waves.

ANS:
The Moho, the inner core, and the shadow zone were discovered by studying seismic waves.

DIF: I   OBJ: 5-4.1

23. What is the shadow zone?

ANS:
The shadow zone is an area on the Earth's surface that does not receive seismic waves from a particular earthquake.

DIF: I   OBJ: 5-4.1

Copyright © by Holt, Rinehart and Winston. All rights reserved.

24. What does the shadow zone tell scientists about Earth's interior?

ANS:
The shadow zone tells scientists that at least part of the Earth's core is liquid.

DIF: I   OBJ: 5-4.1

25. How do "sunquakes" compare with earthquakes?

ANS:
"Sunquakes" are much stronger than earthquakes.

DIF: I   OBJ: 5-4.2

26. What is the relationship between the strength of earthquakes and earthquake frequency?

ANS:
With each step down in earthquake magnitude, the number of earthquakes per year is greater.

DIF: I   OBJ: 5-3.2

27. You learned earlier that if you are in a car during an earthquake and are out in the open, it is best to stay in the car. Briefly describe a situation in which you might want to leave a car during an earthquake.

ANS:
Answers will vary. Students may consider leaving a car to avoid impending danger, such as the car falling off a cliff or being crushed by a tall, heavy object nearby.

DIF: I   OBJ: 5-3.4

28. How did seismologists determine that the outer core of the Earth is liquid?

ANS:
Answers may vary slightly. Seismologists discovered that no seismic waves were detected in the shadow zone. If the outer core were not liquid, seismic waves would not change direction they way they do when they encounter the core. S waves would pass through the core, and P waves would not change direction as drastically as they do. If the outer core were solid, P waves and S waves would be detected in the shadow zone.

DIF: I   OBJ: 5-4.1

Copyright © by Holt, Rinehart and Winston. All rights reserved.

29. Use the following terms to create a concept map: *focus, epicenter, earthquake start time, seismic waves, P waves, S waves.*

ANS:

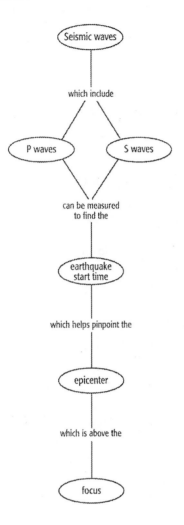

Seismic waves

which include

P waves   S waves

can be measured
to find the

earthquake
start time

which helps pinpoint the

epicenter

which is above the

focus

DIF: II   OBJ: 5-2.2

30. Why do strong earthquakes occur where there have not been many recent earthquakes? (Hint: Think about what gradually happens to rock before an earthquake occurs.)

ANS:
Strong earthquakes occur where there have not been many recent earthquakes because a lot of elastic deformation builds up along active faults where rock has not moved for awhile. The more deformation that builds up, the more energy the rock releases when it finally slips along the fault.

DIF: II   OBJ: 5-1.1

Copyright © by Holt, Rinehart and Winston. All rights reserved.

31. What could be done to solve the wind problem with the seismograph on Mars? Explain how you would set up the seismograph.

ANS:
Answers will vary. One solution might be to place the seismograph in a hole or depression that is shielded from wind.

DIF: II   OBJ: 5-4.2

32. Based on the relationship between earthquake magnitude and frequency, if 150 earthquakes with a magnitude of 2 occur in your area this year, about how many earthquakes with a magnitude of 4 should occur in your area this year?

ANS:
One or two earthquakes with a magnitude of four should occur in my area this year.

DIF: II   OBJ: 5-3.2

The graph below illustrates the relationship between earthquake magnitude and the height of the tracings on a seismogram. Charles Richter initially formed his magnitude scale by comparing the heights of seismogram readings for different earthquakes. Study the graph, and then answer the questions that follow.

Seismogram Height vs. Earthquake Magnitude

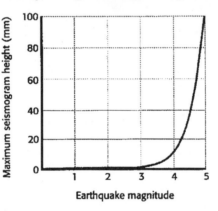

33. What would the magnitude of an earthquake be if the height of its seismogram readings were 10 mm?

ANS:
It would have a magnitude of 4.

DIF: II   OBJ: 5-3.2

Holt Science and Technology
Copyright © by Holt, Rinehart and Winston. All rights reserved.

34. Look at the shape of the curve on the graph. What does this tell you about the relationship between seismogram heights and earthquake magnitudes? Explain.

ANS:
Answers will vary slightly. The relationship is logarithmic, not linear. Students should recognize that seismogram heights increase at a larger rate with each increase in earthquake magnitude.

DIF: II   OBJ: 5-3.2

35. List two things seismologists can learn by studying seismic waves.

ANS:
Sample answers: By studying seismic waves, seismologists can learn the strength of an earthquake, locate an earthquake's epicenter and focus, and draw conclusions about the internal structure of the Earth.

DIF: I   OBJ: 5-4.1

36. How are P waves different from S waves?

ANS:
P waves behave like springs, squeezing and stretching in a straight line. S waves move with a shearing motion, rocking from side to side. Also, P waves are faster than S waves. Finally, P waves can travel through liquid, which S waves cannot do.

DIF: I   OBJ: 5-1.3

37. In an earthquake, opposite sides of a fault suddenly slide past each other, whereas during a nuclear explosion, rocks compress symmetrically in all directions. How might a seismologist be able to tell an earthquake from an underground nuclear explosion?

ANS:
Sample answer: Due to their shearing motion, earthquakes generate much stronger S waves than explosions do. Also, earthquakes last longer than explosions do.

DIF: II   OBJ: 5-1.3

38. In the Earth's inner core, which has a radius of about 1,200 km, S waves travel at 3.5 km/s. About how many minutes will it take an S wave to pass completely through Earth's inner core? Show your work.

ANS:
The diameter of Earth's inner core = 2 × radius = 2 × (1,200 km) = 2,400 km
2,400 km ÷ 3.5 km/s ≅ 686 s
686 s ÷ 60 s/min ≅ 11 minutes

DIF: II   OBJ: 5-1.3

Copyright © by Holt, Rinehart and Winston. All rights reserved.

The graph below compares seismic velocity with depth for S and P waves.

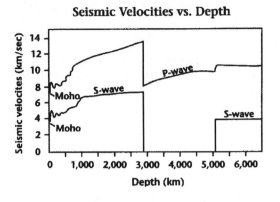

Seismic Velocities vs. Depth

39. At the depth of 2,500 km, about how fast is the S wave traveling?

    ANS:
    At 2,500 km, the S wave is traveling about 7 km/s.

    DIF: II   OBJ: 5-1.3

40. At what range of depths are S waves not transmitted? Explain this finding.

    ANS:
    S waves are not transmitted from a depth of about 2,900 km to a depth of about 5,100 km, which represents the size of the Earth's outer core. S waves do not travel through liquids, such as the outer core.

    DIF: II   OBJ: 5-1.3

Copyright © by Holt, Rinehart and Winston. All rights reserved.

41. Use the following terms to complete the concept map below: *body waves, focus, P waves, epicenter, S waves, seismic waves, surface waves.*

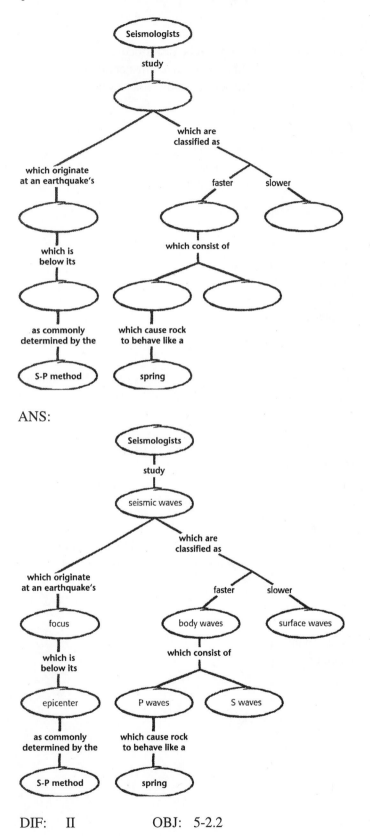

ANS:

DIF: II   OBJ: 5-2.2

Holt Science and Technology
Copyright © by Holt, Rinehart, and Winston. All rights reserved.

MULTIPLE CHOICE

1. The type of magma that often produces a violent eruption can be described as
   a. thin due to high silica content.   c. thin due to low silica content.
   b. thick due to high silica content.   d. thick due to low silica content.

   ANS: B   DIF: I   OBJ: 6-1.2

2. When lava hardens quickly to form ropy formations, it is called
   a. aa lava.   c. pillow lava.
   b. pahoehoe lava.   d. blocky lava.

   ANS: B   DIF: I   OBJ: 6-1.3

3. Volcanic dust and ash can remain in the atmosphere for months or years, causing
   a. decreased solar reflection and higher temperatures.
   b. increased solar reflection and lower temperatures.
   c. decreased solar reflection and lower temperatures.
   d. increased solar reflection and higher temperatures.

   ANS: B   DIF: I   OBJ: 6-2.1

4. Mount St. Helens, in Washington, covered the city of Spokane with tons of ash. Its eruption
   would most likely be described as
   a. nonexplosive, producing lava.
   b. explosive, producing lava.
   c. nonexplosive, producing pyroclastic material.
   d. explosive, producing pyroclastic material.

   ANS: D   DIF: I   OBJ: 6-1.2

5. Magma forms within the mantle most often as a result of
   a. high temperature and high pressure.   c. low temperature and low pressure.
   b. high temperature and low pressure.   d. low temperature and high pressure.

   ANS: B   DIF: I   OBJ: 6-1.2

6. At divergent plate boundaries,
   a. heat from the Earth's core produces mantle plumes.
   b. oceanic plates sink, causing magma to form.
   c. tectonic plates move apart.
   d. hot spots produce volcanoes.

   ANS: C   DIF: I   OBJ: 6-3.2

7. A theory that helps to explain the causes of both earthquakes and volcanoes is the theory of
   a. pyroclastics.
   b. plate tectonics.
   c. climatic fluctuation.
   d. mantle plumes.

   ANS: B   DIF: I   OBJ: 6-3.2

8. Where are volcanoes most likely to form?
   a. near the center of continents
   b. in deep canyons
   c. along plate boundaries
   d. in mountainous areas

   ANS: C   DIF: I   OBJ: 6-3.2

9. Lava that is very runny probably
   a. has a low silica content.
   b. is hotter than most lava.
   c. has been cooled below the surface.
   d. comes from explosive volcanoes.

   ANS: A   DIF: I   OBJ: 6-1.2

10. Volcanic activity is common along the Mid-Atlantic Ridge. This activity occurs at a
    a. mantle plume.
    b. subducted plate.
    c. divergent boundary.
    d. break in the continental crust.

    ANS: C   DIF: I   OBJ: 6-3.2

11. As you move northwest through the Hawaiian Islands, the age of the islands increases. The older islands are northwest of the newer islands because they
    a. ages faster than those to the east.
    b. formed first and were moved away from the hot spot.
    c. were made of less lava than the southeastern islands.
    d. formed from old, recycled, silica-rich magmas.

    ANS: B   DIF: I   OBJ: 6-3.2

12. Which of the following is NOT considered when predicting volcanic eruptions?
    a. seismic activity
    b. atmospheric humidity
    c. internal temperature
    d. changes in the cone's shape

    ANS: B   DIF: I   OBJ: 6-3.3

13. Lava fountains are _____ eruptions that can spray lava into the air.
    a. explosive
    b. nonexplosive
    c. implosive
    d. pyroclastic

    ANS: B   DIF: I   OBJ: 6-1.1

14. In a(n) _____ eruption, a continuous stream of lava pours quietly from the crater of a volcano.
    a. explosive
    b. implosive
    c. nonexplosive
    d. pyroclastic

    ANS: C   DIF: I   OBJ: 6-1.1

15. ____ eruptions often pose a greater threat to property than to human life.
    a. Nonexplosive   c. Implosive
    b. Explosive   d. Pyroclastic

    ANS: A   DIF: I   OBJ: 6-1.1

16. Which of the following eruptions are characterized by relatively calm outpourings of lava?
    a. implosive eruptions   c. explosive eruptions
    b. nonexplosive eruptions   d. pyroclastic eruptions

    ANS: B   DIF: I   OBJ: 6-1.1

17. Which of the following eruptions may resemble a nuclear explosion?
    a. fissure eruptions   c. explosive eruptions
    b. nonexplosive eruptions   d. implosive eruptions

    ANS: C   DIF: I   OBJ: 6-1.1

18. In an eruption in 1915, Mount Lassen in northern California blasted a boulder larger than a
    grown man a distance of 5 km! This was a(n)
    a. lava flow.   c. fissure eruption.
    b. nonexplosive eruption.   d. explosive eruption.

    ANS: D   DIF: I   OBJ: 6-1.1

19. Which of the following eruptions are characterized by blasting both molten and solid rock from a
    volcano?
    a. explosive eruptions   c. lava flows
    b. nonexplosive eruptions   d. fissure eruptions

    ANS: A   DIF: I   OBJ: 6-1.1

20. A volcano is more likely to erupt nonexplosively if its magma
    a. has a high water content.   c. has a low silica content.
    b. has a high carbon dioxide content.   d. All of the above

    ANS: C   DIF: I   OBJ: 6-2.2

21. Magma that has a high silica content
    a. has a thin, runny consistency.   c. has a thick, runny consistency.
    b. has a thick, stiff consistency.   d. has a thin, stiff consistency.

    ANS: B   DIF: I   OBJ: 6-2.2

22. Gases escape from a low-silica magma ____ from a high-silica magma.
    a. more easily than   c. less easily than
    b. at the same rate as   d. None of the above

    ANS: A   DIF: I   OBJ: 6-2.2

Copyright © by Holt, Rinehart and Winston. All rights reserved.

23. A silica-rich magma tends to cause an explosive eruption because it
    a. has a thick, stiff consistency.   c. tends to harden in the volcano's vent.
    b. flows slowly.   d. All of the above

    ANS: D   DIF: I   OBJ: 6-2.2

24. Which lava is so thick in consistency that it barely creeps along the ground?
    a. pahoehoe   c. blocky lava
    b. aa   d. pillow lava

    ANS: C   DIF: I   OBJ: 6-1.3

25. Which lava flows slowly, forming a glassy surface with rounded wrinkles?
    a. pahoehoe   c. blocky lava
    b. aa   d. pillow lava

    ANS: A   DIF: I   OBJ: 6-1.3

26. Which lava pours out quickly, forming a brittle crust that is torn into jagged pieces as the molten
    lava underneath continues to move?
    a. pahoehoe   c. blocky lava
    b. aa   d. pillow lava

    ANS: B   DIF: I   OBJ: 6-1.3

27. Which lava erupts underwater, forming rounded lumps?
    a. pahoehoe   c. blocky lava
    b. aa   d. pillow lava

    ANS: D   DIF: I   OBJ: 6-1.3

28. Which pyroclastic material is the largest?
    a. volcanic blocks   c. lapilli
    b. volcanic bombs   d. volcanic ash

    ANS: A   DIF: I   OBJ: 6-1.3

29. Which pyroclastic material consists of large blobs of magma that harden in the air?
    a. volcanic blocks   c. lapilli
    b. volcanic bombs   d. volcanic ash

    ANS: B   DIF: I   OBJ: 6-1.3

30. Which pyroclastic material consists of pebble-like bits of magma that become solid before they
    hit the ground?
    a. volcanic ash   c. lapilli
    b. volcanic bombs   d. volcanic blocks

    ANS: C   DIF: I   OBJ: 6-1.3

31. Which of the following is NOT a type of pyroclastic material?
    a. lapilli
    b. pillow lava
    c. volcanic blocks
    d. volcanic ash

    ANS: B   DIF: I   OBJ: 6-1.3

32. Which pyroclastic material forms when the gases in stiff magma expand rapidly and the walls of the gas bubbles explode into tiny glasslike slivers?
    a. volcanic blocks
    b. volcanic bombs
    c. lapilli
    d. volcanic ash

    ANS: D   DIF: I   OBJ: 6-1.3

33. Which pyroclastic material is the smallest in size?
    a. volcanic blocks
    b. volcanic bombs
    c. lapilli
    d. volcanic ash

    ANS: D   DIF: I   OBJ: 6-1.3

34. Which type of volcanic material would most likely predominate an explosive eruption?
    a. aa
    b. pahoehoe
    c. volcanic ash
    d. pillow lava

    ANS: C   DIF: I   OBJ: 6-1.3

35. Which volcanic material would most likely predominate a nonexplosive eruption?
    a. lapilli
    b. pahoehoe
    c. volcanic ash
    d. volcanic blocks

    ANS: B   DIF: I   OBJ: 6-1.3

36. Which statement best describes the effect of volcanic ash on the environment?
    a. Volcanic ash can cause the average global surface temperature to drop.
    b. Volcanic ash can mix with rainwater and meltwater, flow downhill, and destroy or move objects in its path.
    c. Volcanic ash can smother crops, causing food shortages and loss of livestock.
    d. all of the above

    ANS: D   DIF: I   OBJ: 6-2.1

37. The eruption of Mount Pinatubo in 1991 caused average global temperatures to drop by as much as 0.5°C. What effect could such a shift in temperature have worldwide?
    a. It could cause worldwide famine.
    b. It could cause the polar icecaps to melt.
    c. It could cause wetter, milder summers and longer, harsher winters.
    d. all of the above

    ANS: C   DIF: II   OBJ: 6-2.1

38. ____ are built out of layers of lava from repeated nonexplosive eruptions.
   a. Cinder cone volcanoes
   b. Shield volcanoes
   c. Composite volcanoes
   d. Stratovolcanoes

   ANS: B   DIF: I   OBJ: 6-2.2

39. ____ are small volcanic cones made entirely of pyroclastic material from moderately explosive eruptions.
   a. Cinder cone volcanoes
   b. Composite volcanoes
   c. Shield volcanoes
   d. Stratovolcanoes

   ANS: A   DIF: I   OBJ: 6-2.2

40. ____ are sometimes referred to as stratovolcanoes.
   a. Shield volcanoes
   b. Cinder cone volcanoes
   c. Composite volcanoes
   d. both (a) and (b)

   ANS: C   DIF: I   OBJ: 6-2.2

41. Which of the following is a type of volcano with gently sloping sides?
   a. a cinder cone volcano
   b. a shield volcano
   c. a stratovolcano
   d. a composite volcano

   ANS: B   DIF: I   OBJ: 6-2.2

42. Which of the following is a type of volcano that is formed by runny lava?
   a. a composite volcano
   b. a shield volcano
   c. a cinder cone volcano
   d. a stratovolcano

   ANS: B   DIF: I   OBJ: 6-2.2

43. Which of the following is a type of volcano that is made entirely of pyroclastic material?
   a. a cinder cone volcano
   b. a stratovolcano
   c. a composite volcano
   d. a shield volcano

   ANS: A   DIF: I   OBJ: 6-2.2

44. Which type of volcano erodes the fastest because its pyroclastic particles are not cemented together by lava?
   a. cinder cone volcano
   b. shield volcano
   c. composite volcano
   d. a stratovolcano

   ANS: A   DIF: I   OBJ: 6-2.2

45. Which of the following is a type of volcano that has a steep slope and a narrow base?
   a. a cinder cone volcano
   b. a composite volcano
   c. a shield volcano
   d. a stratovolcano

   ANS: A   DIF: I   OBJ: 6-2.2

46. Which of the following is a type of volcano that has a steep slope and a broad base?
    a. a cinder cone volcano   c. a composite volcano
    b. a shield volcano   d. both (a) and (b)

    ANS: C   DIF: I   OBJ: 6-2.2

47. Which of the following is a funnel-shaped pit that is found at the top of the central vent in most volcanoes?
    a. crater   c. lava plateau
    b. magma chamber   d. caldera

    ANS: A   DIF: I   OBJ: 6-2.2

48. A ____ forms when a magma chamber that supplies material to a volcano empties and its roof collapses, causing the ground to sink and leaving a large circular depression.
    a. crater   c. lapilli
    b. lava plateau   d. caldera

    ANS: D   DIF: I   OBJ: 6-2.2

49. A ____ is formed when magma erupts from long cracks in Earth's crust and spreads evenly over large areas of Earth's surface.
    a. caldera   c. lava plateau
    b. vent   d. crater

    ANS: C   DIF: I   OBJ: 6-2.2

50. The rock of the Earth's mantle is solid because
    a. it is cool.
    b. it not hot enough to melt.
    c. of high pressure from the rock above it.
    d. of the decrease in pressure as it rises to the surface.

    ANS: C   DIF: I   OBJ: 6-3.1

51. Which of the following raises the melting point of most materials?
    a. an increase in temperature   c. an increase in volume
    b. an increase in pressure   d. none of the above

    ANS: B   DIF: I   OBJ: 6-3.1

52. Rock melts and forms magma when
    a. the temperature of the rock increases.   c. the pressure on the rock increases.
    b. the pressure on the rock decreases.   d. Both (a) and (b)

    ANS: D   DIF: I   OBJ: 6-3.1

53. Because the temperature of the mantle is relatively constant, ____ usually causes magma to form.
    a. increased pressure   c. an eruption
    b. decreased pressure   d. an earthquake

    ANS: B   DIF: I   OBJ: 6-3.1

54. Magma rises toward the surface of the Earth because it is _____ the surrounding rock.
    a. denser than
    b. less dense than
    c. as dense as
    d. None of the above

    ANS: B   DIF: I   OBJ: 6-3.1

55. Lassen Peak erupted in 1914 and continued to erupt intermittently until 1917. In 1916, this volcano was probably classified as
    a. extinct.
    b. dormant.
    c. inactive.
    d. active.

    ANS: D   DIF: I   OBJ: 6-3.3

56. Hawaii's Mauna Kea, the tallest mountain on Earth, is an example of a
    a. shield volcano.
    b. cinder cone volcano.
    c. composite volcano.
    d. stratovolcano.

    ANS: A   DIF: I   OBJ: 6-2.2

57. As tectonic plates separate, a deep crack, or _____, forms between the plates.
    a. caldera
    b. subduction zone
    c. rift
    d. hot spot

    ANS: C   DIF: I   OBJ: 6-3.2

58. The volcano Paracutín, in Mexico, is made entirely of pyroclastic material. Paracutín is an example of a
    a. stratovolcano.
    b. composite volcano.
    c. shield volcano.
    d. cinder cone volcano.

    ANS: D   DIF: I   OBJ: 6-2.2

59. The lavas at Hawaii's Kilauea Crater have the highest temperatures measured on Earth's surface—over 1,200°C! Use the formula $°F = \frac{9}{5}°C + 32$ to calculate 1,200°C in degrees Fahrenheit.
    a. 863°F
    b. 1846°F
    c. 2192°F
    d. 3560°F

    ANS: C   DIF: II   OBJ: 6-3.2

60. Long undersea mountain chains that form at divergent boundaries are called
    a. mid-ocean ridges.
    b. subduction zones.
    c. hot spots.
    d. mantle plumes.

    ANS: A   DIF: II   OBJ: 6-3.2

61. The place where two tectonic plates collide is called a
    a. convergent boundary.
    b. divergent boundary.
    c. rift.
    d. mantle plume.

    ANS: A   DIF: I   OBJ: 6-3.2

Copyright © by Holt, Rinehart and Winston. All rights reserved.

62. Places on the Earth's surface that are located directly above mantle plumes are called
    a.   mid-ocean ridges.   c.   hot spots.
    b.   subduction zones.   d.   lava plateaus.

    ANS: C   DIF: II   OBJ: 6-3.2

63. Scientists think that a combination of heat conducted upward from the core and heat from radioactive elements keeps _____ rising in columns below hot spots.
    a.   mid-ocean ridges   c.   rifts
    b.   lapilli   d.   mantle plumes

    ANS: D   DIF: II   OBJ: 6-3.2

64. Where would you expect to find a long chain of volcanoes?.
    a.   above subducting plates   c.   at convergent boundaries
    b.   above hot spots   d.   all of the above

    ANS: D   DIF: II   OBJ: 6-3.2

Below is a cutaway view of the oceanic crust below the Hawaiian islands. Examine the diagram, and answer the question that follows.

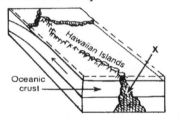

65. The area labeled **X** is located over
    a.   a caldera.   c.   a mid-ocean ridge.
    b.   a hot spot.   d.   a lava plateau.

    ANS: B   DIF: II   OBJ: 6-3.2

66. Magma forms at a convergent boundary when the oceanic plate
    a.   moves downward.   c.   cracks.
    b.   moves upward.   d.   stretches.

    ANS: A   DIF: II   OBJ: 6-3.1

67. Which type of lava does NOT form underwater?
    a.   pahoehoe   c.   blocky lava
    b.   aa   d.   all of the above

    ANS: D   DIF: II   OBJ: 6-1.3

68. Which of the following is a type of volcano that is formed by explosive eruptions of pyroclastic material followed by quieter outpourings of lava?
   a. cinder cone volcano
   b. shield volcano
   c. composite volcano
   d. both (a) and (b)

   ANS: C   DIF: I   OBJ: 6-2.2

COMPLETION

1. A _____ results when an empty magma chamber collapses. (caldera or crater)

   ANS: caldera   DIF: I   OBJ: 6-2.2

2. _____ takes many forms, including volcanic bombs, volcanic blocks, and ash. (Magma or Pyroclastic material)

   ANS: Pyroclastic material   DIF: I   OBJ: 6-1.3

3. Volcanoes made from alternating layers of lava and pyroclastic material are called _____ volcanoes. (cinder cone or composite)

   ANS: composite   DIF: I   OBJ: 6-2.2

4. Parts of tectonic plates that are located above magma plumes are called _____. (hot spots or rifts)

   ANS: hot spots   DIF: I   OBJ: 6-3.2

5. _____ is found deep within the Earth. (Magma or Pyroclastic material)

   ANS: Magma   DIF: I   OBJ: 6-1.1

6. A _____ is a mountain that forms when molten rock is forced to the Earth's surface.

   ANS: volcano   DIF: I   OBJ: 6-1.1

7. A _____ is a river of red-hot lava.

   ANS: lava flow   DIF: I   OBJ: 6-1.1

8. Mount Lassen was once part of a larger, extinct volcano called Mount Tehama. Mount Tehama collapsed when the main magma chamber emptied, forming a _____.

   ANS: caldera   DIF: II   OBJ: 6-2.2

9. _____ is magma that flows onto the Earth's surface.

   ANS: Lava   DIF: I   OBJ: 6-1.1

Copyright © by Holt, Rinehart and Winston. All rights reserved.

10. Magma rises through holes in the Earth's crust called _____.

    ANS: vents   DIF: I   OBJ: 6-1.2

11. _____ material consists of rock fragments created by explosive volcanic eruptions.

    ANS: Pyroclastic   DIF: I   OBJ: 6-1.3

12. The tectonic plate boundaries surrounding the Pacific Ocean have so many volcanoes that these boundaries together are called the _____.

    ANS: Ring of Fire   DIF: I   OBJ: 6-3.2

13. The movement of one tectonic plate under another is called _____.

    ANS: subduction   DIF: I   OBJ: 6-3.2

14. _____ volcanoes are those that have not erupted in recorded history and probably never will again.

    ANS: Extinct   DIF: I   OBJ: 6-3.3

15. _____ volcanoes are those that are not currently erupting but have erupted at some time in recorded history.

    ANS: Dormant   DIF: I   OBJ: 6-3.3

16. _____ volcanoes are those that are in the process of erupting or that show signs of erupting in the very near future.

    ANS: Active   DIF: I   OBJ: 6-3.3

17. Most active volcanoes produce small _____ as the magma within them moves upward and causes the surrounding rock to shift.

    ANS: earthquakes   DIF: I   OBJ: 6-3.3

18. A _____ is a device that measures earthquakes that may precede a volcano eruption.

    ANS: seismograph   DIF: I   OBJ: 6-3.3

19. Measurements of the steepness, or _____, of a volcano can give scientists clues with which to predict eruptions.

    ANS: slope   DIF: I   OBJ: 6-3.3

20. A _____ is a device that detects small changes in the angle of the slope of a volcano.

   ANS: tiltmeter   DIF: I   OBJ: 6-3.3

21. Measuring the presence of certain volcanic _____, such as sulfur dioxide and carbon dioxide, can help scientists predict eruptions.

   ANS: gases   DIF: I   OBJ: 6-3.3

22. _____ are used to measure changes in a volcano's temperature from orbit.

   ANS: Satellites   DIF: I   OBJ: 6-3.3

23. Most of the lava on Earth's continents erupts from long cracks in the Earth's crust called _____.

   ANS: fissures   DIF: I   OBJ: 6-2.2

24. A volcano is more likely to erupt explosively if its magma has a high water and _____ content

   ANS: silica   DIF: I   OBJ: 6-1.2

SHORT ANSWER

For each pair of terms, explain the difference in their meanings.

1. cinder cone volcano/shield volcano

   ANS:
   A cinder cone volcano forms when pyroclastic material erupts and piles up around the volcanic vent. A shield volcano forms when lava erupts and spreads out over large areas.

   DIF: I   OBJ: 6-2.2

2. vent/rift

   ANS:
   A vent is a hole in the Earth's surface that allows lava or pyroclastic material to erupt. A rift is a long, deep crack in the Earth's surface that forms when tectonic plates separate.

   DIF: I   OBJ: 6-2.2

3. lava/pyroclastic material

ANS:
Lava is mostly liquid and is thin and runny. Lava flows out of a volcanic vent onto the ground. Pyroclastic material is mostly solid rock and is blasted into the air in a violent volcanic eruption.

DIF: I   OBJ: 6-1.3

4. lava/magma

ANS:
Magma is hot, liquid rock material beneath the Earth's surface. Lava is magma that flows out onto the Earth's surface.

DIF: I   OBJ: 6-1.2

5. caldera/crater

ANS:
A caldera forms when the roof of a magma chamber collapses. A crater forms when the material above the main vent of a volcano is blasted out.

DIF: I   OBJ: 6-2.2

6. Is a nonexplosive volcanic eruption more likely to produce lava or pyroclastic material? Explain.

ANS:
A nonexplosive eruption is more likely to produce lava than pyroclastic material because lava is thin and runny compared with pyroclastic material.

DIF: I   OBJ: 6-1.1

7. If a volcano contained magma with small proportions of water and silica, would you predict a nonexplosive eruption or an explosive one? Why?

ANS:
A nonexplosive eruption should result. Water turns to steam, which builds up a great amount of pressure, leading to explosive eruptions. Silica-rich magma is thick, allowing it to trap volcanic gases, such as steam, causing explosive eruptions.

DIF: I   OBJ: 6-1.2

Copyright © by Holt, Rinehart and Winston. All rights reserved.

8. Pyroclastic material is classified primarily by the size of the particles. What is the basis for classifying lava?

ANS:
Lava is classified according to how it flows. Blocky lava is thickest and flows very slowly because it is mostly solidified. Aa lava is still thick but is made of smaller blocks and thus flows faster than blocky lava. Pahoehoe lava is thin and runny, and it flows quickly, forming a ropy texture. Pillow lava forms underwater.

DIF: II   OBJ: 6-1.3

9. Briefly explain why the ash from a volcanic eruption can be hazardous.

ANS:
Answers will vary. Volcanic ash is hazardous when it dams rivers, causing floods, and when it smothers crops, resulting in food shortages.

DIF: I   OBJ: 6-2.1

10. Why do cinder cone volcanoes have narrower bases and steeper sides than shield volcanoes?

ANS:
Shield volcanoes are made of lava flows, which are thin and runny, and spread out over large areas. Cinder cone volcanoes are made of pyroclastic material, which is thick and piles up around the volcano.

DIF: I   OBJ: 6-2.2

11. Briefly describe the difference between a crater and a caldera.

ANS:
A crater forms when the rock around the main vent of a volcano is blasted out in an explosive eruption, forming an inverted cone-shaped depression. Most volcanoes have craters at their summits. A caldera forms when the magma chamber that feeds a volcano empties. When this happens, the roof of the magma chamber collapses, forming a circular depression. Calderas are usually much larger than craters.

DIF: II   OBJ: 6-2.2

12. How does pressure determine whether the mantle is solid or liquid?

ANS:
Where there is enough pressure on the mantle, the atoms in the rock are forced to stay close together, keeping it solid. Where this pressure is released, mantle rock melts.

DIF: I   OBJ: 6-3.1

Copyright © by Holt, Rinehart and Winston. All rights reserved.

13. Describe a technology scientists use to predict volcanic eruptions.

ANS:
Answers will vary but should include a discussion of one of the following: measuring changes in the frequency of small earthquakes near the volcano; measuring changes in the slope of the volcano; measuring changes in the ratios of different volcanic gases over time; and measuring changes in how much thermal energy escapes a volcano by using infrared satellite images.

DIF: I   OBJ: 6-3.3

14. Describe the lava flow from a nonexplosive eruption.

ANS:
The lava is a calm stream that can flow for hundreds of kilometers.

DIF: I   OBJ: 6-1.1

15. Describe an explosive eruption.

ANS:
Ash, hot debris, gases, and chunks of rock spew from the volcano.

DIF: I   OBJ: 6-1.1

16. Define blocky lava, pahoehoe lava, and aa lava.

ANS:
Blocky lava is cool, stiff lava that doesn't travel very fast. Pahoehoe lava flows quickly, forms a wrinkled surface, and looks like coiled rope. Aa lava flows slowly and forms a brittle, jagged crust.

DIF: I   OBJ: 6-1.3

17. Describe the shapes of shield, cinder cone, and composite volcanoes.

ANS:
A shield volcano has a broad area with gentle shallow slopes. A cinder cone volcano is generally smaller and steeper, with more angles and sides. A composite volcano is high, covers less area than shield volcanoes, and has sides that become steeper as they near the crater.

DIF: I   OBJ: 6-2.2

18. What causes a caldera?

ANS:
A volcano's magma chamber empties, causing the ground above it to collapse.

DIF: I   OBJ: 6-2.2

Copyright © by Holt, Rinehart and Winston. All rights reserved.

19. What is a lava plateau?

    ANS:
    A lava plateau is a layered formation caused when lava erupts form long cracks, or fissures, and covers a wide area.

    DIF: I   OBJ: 6-2.2

20. What conditions make magma rise?

    ANS:
    When magma is less dense than the surrounding rock and when it has a conduit to move up through, it will rise.

    DIF: I   OBJ: 6-3.1

21. Define a rift.

    ANS:
    A rift is a series of deep cracks that occur where tectonic plates separate.

    DIF: I   OBJ: 6-3.2

22. Briefly describe two methods that scientists use to predict volcanic eruptions.

    ANS:
    Answers will vary but should include two of the following: measuring changes in the frequency of small earthquakes near the volcano; measuring changes in the slope of the volcano; measuring changes in the ratios of different volcanic gases over time; and measuring changes in how much thermal energy escapes a volcano by using infrared satellite images.

    DIF: I   OBJ: 6-3.3

23. Describe how differences in magma affect volcanic eruptions.

    ANS:
    Magma that has a high water and silica content will more likely produce a violent volcanic eruption than magma that has a low water and silica content. Water turns to steam, which builds up a great amount of pressure, which leads to explosive eruptions. Silica makes magma thick, allowing it to trap volcanic gases such as water (steam).

    DIF: I   OBJ: 6-1.2

Holt Science and Technology
Copyright © by Holt, Rinehart and Winston. All rights reserved.
133

24. Along what types of tectonic plate boundaries are volcanoes generally found? Why?

ANS:
Volcanoes are generally found along convergent boundaries because in a subduction zone, oceanic crust is forced downward, which adds water to the mantle. The addition of water lowers the mantle rock's melting point. This melted mantle becomes magma that rises to the surface to form volcanoes.

DIF: I   OBJ: 6-3.2

25. Describe the characteristics of the three types of volcanic mountains.

ANS:
Cinder cones are made from a pyroclastic eruption, are small, and have steep sides. Shield volcanoes are made of lava that runs over great distances before it solidifies, making very large, gently sloped volcanoes. Composite volcanoes are made of both lava and pyroclastic material. Composite volcanoes have large, gently sloping bases and steep sides.

DIF: I   OBJ: 6-2.2

26. Use the following terms to create a concept map: *volcanoes, pyroclastic material, lava, composite volcanoes, shield volcanoes, cinder cone volcanoes.*

ANS:

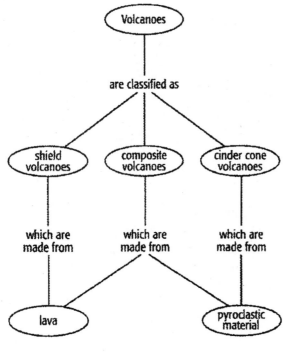

DIF: I   OBJ: 6-2.2

Holt Science and Technology
Copyright © by Holt, Rinehart and Winston. All rights reserved.

27. Imagine that you are exploring a volcano that has been dormant for some time. You begin to keep notes on the types of volcanic debris you encounter as you walk. Your first notes describe volcanic ash, and later your notes describe lapilli. In what direction would you most likely be traveling—toward or away from the crater? Explain.

ANS:
You would be traveling toward the volcano because the larger the pyroclastic material is, the closer it will be to the vent. It takes more energy to move larger particles than it does to move smaller particles.

DIF: II   OBJ: 6-1.3

28. Loihi is a future Hawaiian island in the process of forming on the ocean floor. Considering how this island chain formed, tell where you think the new volcanic island will be located and why.

ANS:
The new island will be located southeast of Hawaii because the Pacific plate is moving toward the northwest.

DIF: II   OBJ: 6-3.2

29. What do you think would happen to the Earth's climate if volcanic activity increased to 10 times its current level?

ANS:
The overall surface temperature of the Earth would decrease because the volcanic ash in the atmosphere would block out much of the sun's energy. (Students may also note that volcanic eruptions release large amounts of CO_2, which could cause long-term global warming.)

DIF: II   OBJ: 6-2.1

30. Midway Island is 1,935 km northwest of Hawaii. If the Pacific plate is moving to the northwest at 9 cm per year, how long ago was Midway Island located over the hot spot that formed it?

ANS:
1 km = 1,000 m = 100,000 cm
1,935 km = 193,500,000 cm
193,500,000 ÷ 9 cm per year = 21,500,000 years

DIF: II   OBJ: 6-3.2

The following graph illustrates the average change in temperature above or below normal for a community over several years. Use the graph to answer the questions that follow.

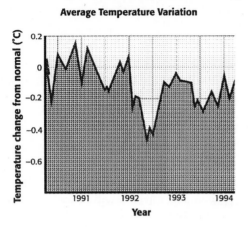

31. If the variation in temperature over the years was influenced by a major volcanic eruption, when did the eruption most likely take place? Explain.

ANS:
The eruption probably happened in 1992 because that year had the lowest temperature below normal. The volcanic ash that was erupted into the atmosphere blocked the sunlight and lowered the temperature.

DIF: II   OBJ: 6-2.1

32. If the temperature were plotted only in yearly intervals rather than several times per year, how might your interpretation be different?

ANS:
If the temperature was measured once a year, the graph would indicate that 1991 had the lowest temperature. This would mean that the eruptions happened in 1991 instead of 1992.

DIF: II   OBJ: 6-2.1

33. How are silica and water vapor related to a volcano's explosiveness?

ANS:
The amount of silica and water vapor in magma determines a volcano's explosiveness. The volcano is more likely to be explosive if its magma contains high amounts of water and silica.

DIF: I   OBJ: 6-1.2

34. The sides of Mount Rainier, in Washington, are covered with snow and glaciers. If Mount Rainier were to erupt explosively, what effects would you expect?

ANS:
Sample answer: The large amounts of hot ash from the eruption could mix with water from melted snow and glaciers, causing dangerous mud flows.

DIF: II   OBJ: 6-2.1

35. After 2,000 years of calm, Krakatoa, a volcano in Indonesia, exploded violently in 1883. Ash flew 80 km into the air and covered an area of 800,000 km^2 with pyroclastic debris. How do you think this mighty eruption affected the global climate?

ANS:
Sample answer: The volcanic ash reached the upper atmosphere and blocked out enough sunlight to cause a decrease in global temperatures.

DIF: II   OBJ: 6-2.1

36. The eruption of Mount Vesuvius in A.D. 79 buried Pompeii's 155 acres under 6 m of volcanic ash. Calculate the approximate volume of volcanic ash that buried Pompeii. Express your answer in m^3. (Hint: 1 acre = 4,047 m^2) Show your work.

ANS:
volume of the ash = 155 acres $\times$ 4,047 m^2/acre $\times$ 6 m
Pompeii was covered in 3,763,710 m^3 of volcanic ash.

DIF: II   OBJ: 6-1.3

Copyright © by Holt, Rinehart and Winston. All rights reserved.

The following graphs show seismograph and tiltmeter readings from a certain volcano. The times represented by points A and B are the same for both graphs. Examine the graphs, and answer the questions that follow.

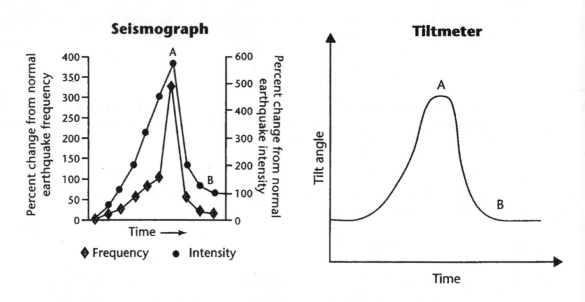

37. What event probably occurred at time A? Explain your reasoning.

ANS:
The volcano probably erupted. The earthquake frequency and intensity both peaked sharply at time A, and the greatest change in the angle of the cone's wall occurred at time A.

DIF: II   OBJ: 6-3.3

38. Why does the angle recorded by the tiltmeter decrease around time B?

ANS:
Time B is after the volcanic eruption. The eruption released the magma that had caused the volcano's surface layers to bulge upward and outward. Thus, the angle measured by the tiltmeter at time B was reduced.

DIF: II   OBJ: 6-3.3

Copyright © by Holt, Rinehart and Winston. All rights reserved.

39. Use the following terms to complete the concept map below: *pyroclastic material, mantle plume, nonexplosive volcanoes, lapilli.*

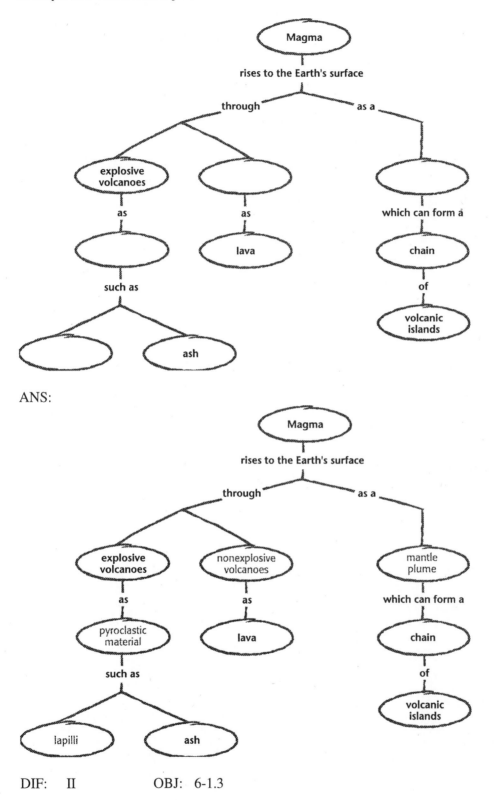

ANS:

DIF: II   OBJ: 6-1.3

Copyright © by Holt, Rinehart and Winston. All rights reserved.